Studies in Computational Intelligence

Volume 871

Series Editor

Janusz Kacprzyk, Polish Academy of Sciences, Warsaw, Poland

The series "Studies in Computational Intelligence" (SCI) publishes new developments and advances in the various areas of computational intelligence—quickly and with a high quality. The intent is to cover the theory, applications, and design methods of computational intelligence, as embedded in the fields of engineering, computer science, physics and life sciences, as well as the methodologies behind them. The series contains monographs, lecture notes and edited volumes in computational intelligence spanning the areas of neural networks, connectionist systems, genetic algorithms, evolutionary computation, artificial intelligence, cellular automata, self-organizing systems, soft computing, fuzzy systems, and hybrid intelligent systems. Of particular value to both the contributors and the readership are the short publication timeframe and the world-wide distribution, which enable both wide and rapid dissemination of research output.

The books of this series are submitted to indexing to Web of Science, EI-Compendex, DBLP, SCOPUS, Google Scholar and Springerlink.

More information about this series at http://www.springer.com/series/7092

Minakhi Rout · Jitendra Kumar Rout ·
Himansu Das
Editors

Nature Inspired Computing
for Data Science

 Springer

Editors
Minakhi Rout
School of Computer Engineering
Kalinga Institute of Industrial Technology
Deemed to be University
Bhubaneswar, Odisha, India

Jitendra Kumar Rout
School of Computer Engineering
Kalinga Institute of Industrial Technology
Deemed to be University
Bhubaneswar, Odisha, India

Himansu Das
School of Computer Engineering
Kalinga Institute of Industrial Technology
Deemed to be University
Bhubaneswar, Odisha, India

ISSN 1860-949X   ISSN 1860-9503   (electronic)
Studies in Computational Intelligence
ISBN 978-3-030-33822-0   ISBN 978-3-030-33820-6   (eBook)
https://doi.org/10.1007/978-3-030-33820-6

This Springer imprint is published by the registered company Springer Nature Switzerland AG
The registered company address is: Gewerbestrasse 11, 6330 Cham, Switzerland

Preface

Nowadays, nature inspired computing has emerged as a new area of research which has taken a high pick in almost every field of computer science as well as data science in specific. Due to high popularity in data science, it attracts researchers from almost every field of data science to use it intelligently to analyse a massive amount of data. Nature inspired computing solves complex real world problems in various environmental situations by observing natural occurring phenomena. It helps to develop various computing techniques such as evolutionary computation, swarm intelligence, artificial immune system, neural network, etc. This is a kind of computing that mimics the behaviours of any biological agents or groups, the way they exchange and process information by doing collaborative task to achieve a particular goal or making a decision. The action of these biological agents motivates and inspires to imitate their activity to solve the problems in various era of data science. Based on the above idea, the researcher starts to use a group of solutions as agents instead of single solution (agent) to solve real-world problems or to achieve the specific goal with less time by covering the entire search space of the problem to get an optimal solution.

In twenty-first century, the data that are being generated at an alarming rate often termed as Big Data. It leads to a crucial task to store, manage and analyse these data to retrieve useful information to take any decision in various fields. So data scientists are looking for efficient, scalable, versatile techniques to analyse these huge amounts of data. Data science is a perfect blend of tools, algorithms and machine learning approaches to get the new insights of the data to discover hidden patterns out of it. Nature inspired computing attracts the data scientists to blend the advantages of it with data science to improve the quality of analysis or outcomes.

The objective of this book is to provide all aspects of computational intelligence based on nature inspired computing and data science to build intelligent systems for real-time data analytics. It includes most recent innovations, trends, practical challenges and advances in the field of nature inspired computing techniques and data science. By looking at its popularity and application in interdisciplinary research fields, this book focuses on the advances in nature inspired computing and its usefulness in data science for analysing the data in various aspects. To achieve the objectives, this book includes 12 chapters contributed by promising authors as follows:

In chapter "An Efficient Classification of Tuberous Sclerosis Disease Using Nature Inspired PSO and ACO Based Optimized Neural Network", Ripon et al. addresses the use of convolutional neural network optimized with nature inspired approaches such as Particle Swarm Optimization and Ant Colony Optimization for classification of tuberous sclerosis disease. In chapter "Mid-Term Home Health Care Planning Problem with Flexible Departing Way for Caregivers", authors focus on home health care centres for planners that make proper decision through managing the depart way of caregivers. It satisfies the real-life constraints and minimizes the total operational cost as well as find the optimal solution efficiency. Chapter "Performance Analysis of NASNet on Unconstrained Ear Recognition" focused on the biometric recognition, through identification of the subject from the ear. Chapter "Optimization of Performance Parameter for Vehicular Ad-hoc NETwork (VANET) Using Swarm Intelligence" addresses Ant Colony Optimization to establish multiple routes between nodes which is vital in the network where network connectivity is random and is frequently changing. This also helps in the parallel transmission of packets in multi-path which reduces end-to-end delay.

In chapter "Development of Fast and Reliable Nature-Inspired Computing for Supervised Learning in High-Dimensional Data", the author presents a fast and reliable nature inspired training method for artificial hydrocarbon networks for handling high-dimensional data. Chapter "Application of Genetic Algorithms for Unit Commitment and Economic Dispatch Problems in Microgrids" presents the Genetic Algorithm (GA) to address the combined problem of unit commitment and economic dispatch. Similarly in chapter "Application of Genetic Algorithms for Designing Micro-Hydro Power Plants in Rural Isolated Areas—A Case Study in San Miguelito, Honduras", the authors address the use of Genetic Algorithm to assist the design of micro-hydro power plants, which finds the most suitable location of the different elements of a micro-hydro power plants to achieve the most efficient use of the resources. Chapter "Performance Evaluation of Different Machine Learning Methods and Deep-Learning Based Convolutional Neural Network for Health Decision Making" presents the comparative analysis of performance among different machine learning algorithms and deep neural networks for diabetes disease prediction. Chapter "Clustering Bank Customer Complaints on Social Media for Analytical CRM via Multi-objective Particle Swarm Optimization" describes the two variants (multi-objective particle swarm optimization along with heuristics of K-means and multi-objective particle swarm optimization along with the heuristics of spherical K-means) of multi-objective clustering algorithm in the applications of CRM banking industry for sentiment analysis. Chapter "Benchmarking Gene Selection Techniques for Prediction of Distinct Carcinoma from Gene Expression Data: A Computational Study" addresses a review on gene selection approaches for simultaneous exploratory analysis of multiple cancer datasets. It provides a brief review of several gene selection algorithms and the principle behind selecting a suitable gene selection algorithm for extracting predictive genes for cancer prediction. Chapter "An Evolutionary Algorithm Based Hybrid Parallel Framework for Asia Foreign Exchange Rate Prediction" provides an evolutionary algorithm based hybrid parallel model to enhance the prediction of Asia foreign exchange rate.

Topics presented in each chapter of this book are unique to this book and are based on unpublished work of contributed authors. In editing this book, we attempted to bring into the discussion all the new trends and experiments that have made on data science using nature inspired computing. We believe this book is ready to serve as a reference for a larger audience such as system architects, practitioners, developers and researchers.

Bhubaneswar, India   Himansu Das

The original version of the book was revised: Volume number has been corrected. The correction to the book is available at https://doi.org/10.1007/978-3-030-33820-6_12

Acknowledgements

The making of this edited book was like a journey that we had undertaken for several months. We wish to express our heartfelt gratitude to our families, friends, colleagues and well-wishers for their constant support throughout this journey. We express our gratitude to all the chapter contributors, who allowed us to quote their remarks and work in this book. In particular, we would like to acknowledge the hard work of authors and their cooperation during the revisions of their chapters. We would also like to acknowledge the valuable comments of the reviewers which have enabled us to select these chapters out of the so many chapters we received and also improve the quality of the chapters. We wish to acknowledge and appreciate the Springer team for their continuous support throughout the entire process of publication. Our gratitude is extended to the readers, who gave us their trust, and we hope this work guides and inspires them.

Contents

An Efficient Classification of Tuberous Sclerosis Disease Using Nature Inspired PSO and ACO Based Optimized Neural Network

**Shamim Ripon, Md. Golam Sarowar, Fahima Qasim
and Shamse Tasnim Cynthia**

Abstract Tuberous sclerosis disease is a multi-system genetic disorder that broadly affects the central nervous system resulting in symptoms including seizures, behavior problems, skin abnormalities, kidney disease etc. This hazardous disease is caused by defects, or mutations of two genes-TSC1 and TSC2. Hence, analysis of TSC1 and TSC2 gene sequences can reveal information which can help fighting against this disease. TSC2 has 45 kilobases of genomic DNA, 41 known exons, and codes for a 5474-base pair transcript. On the other hand, the TSC1 gene spans about 53 kb of genomic DNA with 23 exons coding for hamartin, a hydrophilic protein with 1164 amino acids and 130 kb DNA. It is not possible to manually extract and analyze all the hidden information lies in TSC1 and TSC2 in wet lab. Machine learning approaches have been extensively applied to discover hidden information lies in any dataset. Efficient machine learning approaches need to be discovered to analyze TSC1 and TSC2. This chapter concentrates on using convolutional neural network optimized with nature inspired approaches such as, Particle Swarm Optimization (PSO) and Ant Colony Optimization (ACO). The main challenge of any machine learning approaches is to optimize its parameters effectively. Since, both PSO and ACO are iterative optimization approaches to iteratively develop a candidate solution it can be employed with machine learning approaches to reduce the effort to formulize the parameters of various machine learning techniques. Besides, all the weights, biases, learning rates are optimized with PSO and ACO algorithms. The proposed approach has been implemented for classification of tuberous sclerosis disease. Additionally, for comparison purpose we have employed decision tree, naïve bayes, polynomial regression, logistic regression, support vector machine, random forest etc. Comparative analysis of time and memory requirements of all the approaches have been performed and it is found that the efficiency and time requirements of proposed approach outperform its competitors. Meanwhile, Apriori algorithm have been applied to generate association rules and to extract effective information regarding dependencies of attributes with each other to identify those that are responsible to

S. Ripon (✉) · Md. Golam Sarowar · F. Qasim · S. T. Cynthia
Department of Computer Science and Engineering, East West University, Dhaka, Bangladesh
e-mail: dshr@ewubd.edu

© Springer Nature Switzerland AG 2020
M. Rout et al. (eds.), *Nature Inspired Computing for Data Science*,
Studies in Computational Intelligence 871,
https://doi.org/10.1007/978-3-030-33820-6_1

1

empower this disease. For exploration and extraction of different significant information from the sequence of TSC1 and TSC2 gene, frequent itemset along with analysis of mutation sequence combining other approaches have been illustrated. Statistical analysis on the same dataset reveals the similar findings as the Apriori algorithm.

1 Introduction

Tuberous sclerosis is a rare genetic disorder that affects cell differentiation, expansion, and relocation, resulting in various lesions that can affect every organ system in our body. This disorder results from mutation in the sequence of either TSC1 or TSC2 gene, or both. The turmoil can cause a wide scope of potential signs and indications and it is related to the arrangement of benign (non-cancerous) tumors in different organ frameworks in the body. The most affected parts of the body are skin, brain, eyes, heart, kidneys and lungs. Tuberous sclerosis can cause gentle disease in which individuals go unfamiliar into adulthood or it can cause significant troubles or the turmoil can cause possibly extreme, hazardous complexities.

The signs, indications and seriousness of the disorders can differ significantly from one person to another, even in individuals in the same family. Numerous infants have white fixes or spots (hypomelanotic macules) on their skin during childbirth or ahead of schedule amid early stages. A rash or plaque of abnormal skin usually develops on the lower back and is sometimes referred to as a texturing or feeling of an orange peel, e.g. rough and bumped. It is often used as an uncomplicated, thick and fleece-colored rash that is called the shagreen patch. If one parent has tuberous sclerosis, each child born to that parent has a half shot of acquiring that disease but most of the time, there is no family history. It is a transformation that causes it. Some signs and side effects of tuberous sclerosis occur less often than previously illustrated. These results are portrayed as minor points of disarray. Such indications include patches or territories of aided shadings (retinal colorless fixation) or an absence of shading that influences the retina. Some people can create setting or little openings in the veneer of teeth (dental finish setting) or arrange the sinewy development of the mouth (intraoral fibromas), particularly in gums. In some cases, in the midst of youth, before adulthood or adult people can create confetti skin soreness. These injuries are modest (1–3 mm) and shade detections (hypopigmented) lighter than the skin surrounding them. They may be scattered across specific skin areas.

TSC1 and TSC2 are the genes whose alteration cause the TSC disease. Genes provide guidance on the production of proteins which are essential for various elements in the body. The protein item may be defective, wasteful or missing at the time a transformation of gene occurs. This can affect numerous organ frameworks in the body, including the cerebrum, depending on the particular protein elements. Both genes are coded for the production of proteins of Hamartin and Tuberin. Both of the proteins act as tumor development silencers and furthermore manage cell multiplication and separation. The psychological exposure of TSC are a direct result of the development of hamartia, hamartomas and cancerous hamartoblastomas.

The cause of birth is 1 in 6000 in tuberous sclerosis the most common TSC neurological disorder is the epilepsy, which results from tuberous sclerosis over the years. Around 40,000–80,000 individuals in the United States have tuberous sclerosis. The pervasiveness in Europe is evaluated to be roughly 1 out of 25,000 to 1 out of 11,300. Upwards of 2 million individuals worldwide are accepted to have the turmoil. Males and females are influenced in equivalent numbers and the disorder happens in all races and ethnic groups. Until now this disease is clinically diagnosed through the identification of different mutation types in the gene sequence of TSC1 and TSC2. If a mutation can be identified in the gene sequence, it will be sufficient for the disease to be diagnosed. As the database of mutations increases day by day, clinical intervention alone is not sufficient to detect TSC within a short time. It will be expensive as well as tedious to guarantee of the presence of this disease.

A machine assisted mechanism can defeat the constraints of manual diagnosis of TSC effectively. Be that as it may, till now no such procedure has been utilized to recognize and analyze the disease. To deal with the issue, we have proposed to optimize Convolutional Neural Network to automatically analyze and predict the disease from the available dataset. For the ability to learn from the training data automatically, Convolutional Neural Network has recently been used for better performance. But the principal concern on using the CNN is the period of time it takes to prepare and update the phase of the parameters utilized in CNN. This is where the optimization with different algorithms come in and the integration between them can uncover exceptional execution alongside proficiency while training CNN. The first optimization is done by using Particle Swamp Optimization (PSO) [19] which is an evolutionary optimization technique. Every swarm particle adjusts itself by considering experience of own and companion and determine global best value of any parameter.

The second optimization iss done using Ant Colony Optimization (ACO) [20]. It is a framework dependent on specialists which simulate the natural behavior of ants, including systems of collaboration and adjustment. Each insect actually performs its own task independently of other colonial members in a colony of social insects like the ants, the bees, the wasps and the termites. The task performed by various insects are nevertheless linked so that, through cooperation, the colony can solve any complex issue as a whole. This led researchers to develop an algorithm following the nature of the insects working together in an efficient manner. We have conducted this Ant Colony Optimization with the Convolutional Neural Network for making this process more efficient and reducing the total time consumption. While training the dataset with CNN only, the back propagation of this algorithm usually takes long time to compute because after calculating the error rate we need to go for the derivatives to get the optimal solution. This becomes harder when the number of layers increases. So to get rid of this problem we have integrated Ant Colony Optimization with Convolutional Neural Network without computing back propagation and through this procedure, we can also get the solution weights and bias values.

We have also analyzed the mutations dataset using data mining techniques for finding the relations among the attributes of mutations sequences. We have used Apriori and decision tree rule induction mechanism for generation of association

rules which can lead experts to infer critical decisions depending on the association rules. Data mining techniques are basically ensemble types machine algorithms
which extracts significant information within huge datasets. Since, our considered
Tuberous Sclerosis dataset is comparatively large with more than 30,000 mutations
sample along with their position, length, Alleles, Origin of Alleles, Significance
and so we have implemented data mining approaches to provide clear estimations,
principles and information regarding Tuberous Sclerosis disease. Nowadays, various
data mining as well as machine learning techniques [21, 22] have a huge impact on
medical science [24]. They are able to generate classification rules by formed of IF
condition and THEN statement.

The rest of the chapter is organized as follows. Review of similar works are
briefly explained in Sect. 2. An overview of the dataset along with some statistical
analysis of the attributed are illustrated in Sect. 3. Section 3 describes the proposed
methods and explains the techniques that have been applied in this chapter. Analysis
of results obtained from the experiments are demonstrated in Sect. 4. Both individual
and hybrid algorithms are explained in this section. Section 5 shows the analysis of
obtained results. Finally, Sect. 6 concludes the paper by summarizing the chapter
and then outlining the future plan after this work.

2 Literature Review

There are several contributions to the prediction and classification of Tuberous Sclerosis Complex disease, however most of the works were done traditionally. But,
determination of TSC in a traditional way is not sufficiently workable in this era
of technology, big data and science. The authors of contribution [1] however, state
that, TSC manifestations differ dramatically from one person to another, and even
between similar twins [2]. According to their research, remarkable improvements
have been made in the comprehension and analysis of the functions TSC1 and TSC2
also the cellular and molecular implications of the loss of functional mutation in
these genes. The most prominent findings of their study is to identify core failures
that covered all key elements of this disease including the annotations and distribution of bio-specimens; an improved animal model; and a wider network of clinical
trials which includes non-neurological TSC manifestations.

Moreover, the authors in the work [3] state that, according to clinical or genetic
criteria, they enrolled 130 infants with a certain TSC which followed up to the age of
36 months. They have frequently examined and identified histories of medical and
seizure, physical, neurological and developmental examinations. Also, they evaluated the age groups in which TSC's major and minor characteristics were identified.
However, they found TSC's most frequent earlier characteristics which are cardiac
rhabdomyomas (59%) and hypomelanotic macules (39%). Some common diagnostic
characteristics such as hypomelanotic macules (94%), tubers or other cortical dys-
plasias (94%) have also been found. However, here the entire workflow is manual
and constant monitoring is mandatory which requires a lot of time. Whereas, we have

proposed a machine learning approach which includes CNN optimized with ACO and CNN optimized with PSO [23] that understand better accuracy and efficiency in the earliest stage of detection of this disease. Especially in the case of big data [25], our depicted accuracies are much higher and more comprehensible. In a matter of a few second we can now detect and predict the disease rather than wait for 36 months.

According to the authors of the contribution [4], a preliminary review of suspected TSC children has been shown due to single or multiple heart tumors and medical reports were also observed depicting TSC signs and symptoms. 82/100 children were diagnosed with TSC and early TSC signs have been observed during the first four months of their life. However, the most helpful clinical studies were MRI, skin testing and echocardiography for early TSC diagnosis. Meanwhile, our proposed mechanism focuses on automated detection by analyzing the basic level of protein sequence that is primarily responsible for this disease. Moreover, there are lots of approaches applied for mutation analysis of the TSC1 and TSC2 genes. In the work [5], TSC1 and TSC2 gene mutation analysis was conducted in 490 patients with TSC diagnosis. From the authors' findings, minimal clinical data were obtained in 276 cases to achieve a definite diagnosis of TSC based on Roach et al. [6] criteria. However, the 291 clinically informed patients were 0–60 years of age, with a median age of 13.0 at the time of referral. The authors also analyzed correlation between different clinical features in the 276 definitive TSC patients. Most patients were mentally retarded 166 among 276 (60%) and seizures occurred in 163 of these patients (98%). Conversely, only 37 out of the 63 mentally disrupted patients had seizures (59%). They also found the patients with (sub)-cortical tubers had seizures (86%) and 56% of patients had seizures without tubers. Moreover, the authors identified pathogenic mutations in 362 patients where they found 82 TSC1 mutations and 280 TSC2 mutations.

According to the work in [7], 470 TSC patients were examined for mutations in SSCP in combination with TSC1 and TSC2. The authors detected two sequence changes 5440delTG and 5438delTGTG which involves the TSC2 stop codon. From their research on a two generation Danish family, they found index patient 2057 who has facial angiofibroma, ungual fibroma, hypomelanotic macules, Subependymal nodules and cortical tubers actually suffers from epilepsy and has mild mental disorder. Moreover, individual 2055 has facial angiofibroma, ungual fibroma and hypomelanotic macules whereas individual 2044 suffers from facial angiofibroma, ungual fibroma, hypomelanotic macules and Subependymal nodules. However, the other family members were not classified as clinically affected but in case of individual 1494, who was too young for the definitive diagnosis but no TSC were detected at this case. They also investigated the effects from these two sequence changes on the functions of tuberin which is a TSC2 gene product.

Another journal article [8], where the authors investigated the effects of TSC1 variants in the individuals with signs of TSC. They examined on a two generation of a family who has TSC and detected TSC1 c.350T > C change. They found that the index patient (I:1) was affected by epilepsy since the age of 22 with a certain TSC diagnostic criteria which includes facial angiofibroma, ungual fibroma, hypomelanotic macules, ashagreen patch and cerebral white matter migration lines. Moreover, c.350T > C change has been detected in the individual (II:2) who is the youngest child of the family. This child grows normally until the age of 4, despite the fact that the child was somewhat hyperactive. The child is now completely paralyzed. However, the eldest child individual (II:1) had not yet tested for the c.350T > C change but suffered severely anoxia at birth and has serious infantile spastic tetraplegia and epilepsy encephalopathy. The authors examined quite a few more families with two or more generations and found several TSC1 variants changes.

Many researches have done research on approaches to analyze protein sequences as well. The authors of the contribution [9] asserted that, methods for separating and isolating proteins are available. They have focused on automated learning approaches such as Edman degradation [10] and Phenyl Isothiocyanate degradation (PITC) [11]. They focused mainly on the Edman degradation based on the PITC reactant with the protein α-amino group. The degradation was extensively applied in the structural analysis of proteins and limitations were found such as most degradations were restricted to a few cycles and therefore compositional and partial sequence analysis were used. In our work, we concentrated mainly on CNN optimized with PSO [28] and ACO, which show higher precision and lowest memory consumption.

However, in another work [12] which is related to protein sequence analysis illustrates that applied tandem mass spectra [13] which is an automated machine learning approach for protein sequence quantitative analysis. They changed the method by which data is obtained in proteomic analyses by shotgun. They also analyzed the soluble fraction of whole cell lysates from yeast, metabolizing in vivo 15 N, and compared the quantitative merit figures for this method.

Furthermore, in the work [14], the authors focused primarily on epilepsy and autism treatment approaches in TSC patients. For the treatment of these diseases, they used manual approaches such as tubers, molecular genetics and animal models. Whereas, our proposed efficient machine learning approaches usually detect TSC disease in the human body. Since our dataset is very large, traditional or manual detection is impossible.

According to the authors of the contribution [15], in order to provide appropriate treatment, early diagnosis of TSC is very important. Fetal ultrasounds and MRI imaging techniques were used in order to capture previous TSC-related lesions like cardiac rabdomyomas, Subependymal nodules, cortical tubers and renal cysts. According to the work [16], the authors performed analysis of ictal and interictal EEG of four patients with TSC and they were undergoing surgery for epilepsy. After that, these recordings were submitted to independent component analysis (ICA) and they used sLORETA algorithm for this. All four patients had long-term video-EEG recordings and there was 256 Hz of sampling and a band rate of 0.1–70 Hz. For patients 1 and 3, the electrode positions were measured by the DeMunck et al. method also the standard positions for the 10–20 system for patients 2 and 4 were obtained. From their research, they found that all the patients had multifocal interictal spikes in the scalp EEG but there was always a grossly imbalanced type present including consistent morphology and dipolar topography. However, the sources from these interictal spikes have been located in the lobe of a fastest growing tuber for all patients with the maximum sLORETA statistical score close to the linkage between the brain tissue

and the dysplastic brain. Whereas, other topographies of spikes have been correlated with very different locations of their generators, which generally provide little
coherence in defining the brain area of the epileptical patient from interictal records.

In [17], the authors reported a new approach for the detection of mutations in TSC
such as: a denaturing gradient gel electrophoresis (DGGE) analysis for small TSC2
mutations, a multiplex ligation-dependent probe amplification (MLPA) analysis for
large deletions and duplications in TSC1 or TSC2, and a long-range PCR/sequencing-
based analysis for small TSC1 mutations. From their research, 65 clinically diagnosed
Danish patients were analyzed for TSC and 51 patients were identified with disease
mutations (78%). They found 36 small TSC2 mutations, 4 large deletions with TSC2
mutations and 11 small TSC1 mutations. However, of the small mutations, twenty-
eight are novel. Moreover, they have also developed a functional test for the missense
mutations to illustrate that the mutations affect the function of TSC2.

3 Dataset Overview

Dataset for "Tuberous Sclerosis Complex" disease has been extracted from National
Centre for Biotechnology Information. Single nucleotide polymorphisms (SNPs)
have been considered and assembled together for TSC1 and TSC2 proteins in the
dataset to analyze and classify tuberous sclerosis disease. The dataset contains a
total of 31,000 patients and this disorder affects all patients. Each patient has individual records in the dataset which includes SNP sequence (nominal value type),
position of mutation in the sequence (Position: discrete numeric type), length of the
SNP sequence (Length: discrete numeric type), Mutated base pair (Alleles: nominal
value), Origin of the alleles (Allel_origin: nominal value) and finally Significance
(nominal value type). Among these attributes first five attributes are referred as input
attributes and significance is considered as target attribute or classification attribute.
Four of the collected attributes are categorical, and two are numerical. There are five
class labels in classification attribute. They are,

- BENIGN—which means the disease is not harmful in effect
- PATHOGENIC—which means the patient is in the process of being infectious by
  tuberous sclerosis
- MALIGNANT—which means the disease is in the infectious phase
- OTHER—rather than those three the disease is classified as other
- UNTESTED—which means the disease is not identified yet

Figure 1 represents a sample instances and features of our dataset.

Table 1 illustrates the position and length of two distinct numerical attributes
and their statistical information. We have analyzed some statistical analysis over the
dataset. 50% of the whole dataset is classified as Benign among 31000 significance
in the dataset. 30% is subsequently classified as malignant. In addition, 11, 5, and
4%, are classified as pathogenic, untested and other. Thus, the disease is not harmful
and effective for 50% of patients, while 30% are likely to become infected with it.

Sequence	Position	Length	Alleles	Allel_Origin	Significance
CTCTGAAGGTCCAAAGAGTTTCT....	501	1001	C/T	T/C	Benign
ACGGAACCAGGAAACTAGACTGT....	501	1001	A/G	G/A	Malignant
CTTTACTGTAAGGGTGTGACAGA....	501	1001	A/G	G/A	Pathogenic
ACGGAACCAGGAAACTAGACTGT....	501	1001	A/C/G/T	T/G/C/A	Benign
ACAAAGCTGAATTAAATGTGTGG....	501	1001	A/C/T	T/C/A	Benign
GATAGTAGGTGGCAATCTTAAGT....	401	801	A/G	G/A	other
CCAGGAGTTAGAGACCAGCCTGA....	501	1001	C/T	T/C	Malignant
AAGGATTCAGTCATAAAGAGGTA....	501	1001	A/G	G/A	Benign
AGGTGAACTGTTACGACCCAGTA....	501	1001	A/G	G/A	Pathogenic
AAGGATTCAGTCATAAAGAGGAT....	501	1001	A/G	G/A	other
CTCTGAAGGTCCAAAGAGTTTAG....	501	1001	A/G	G/A	Benign
AAGGATTCAGTCATAAAGAGGA....	501	1001	A/G	G/A	Malignant
CTTTACTGTAAGGGTGTGACAGA...	501	1001	C/T	T/C	Benign
AGGTGAACTGTTACGACCCAGG....	501	1001	A/G	G/A	Pathogenic
CTCTGAAGGTCCAAAGAGTTTC....	501	1001	A/G	G/A	Benign

Fig. 1 Sample instances and features in the dataset

Table 1 Analysis of numerical attributes

Attributes name	Minimum	Maximum	Mean	Standard deviation
Position	26	501	52.93916	7.30150
Length	51	1001	105.4053	10.302

However, each significance and their appearances in the dataset have been illustrated in Fig. 2.

Moreover, About 785 alleles in the dataset are unique among the 31000 alleles. However, only few of them actually affect the spread of this disease. We considered

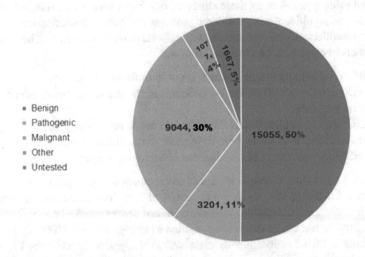

Fig. 2 Total occurrence of each significance in dataset

Table 2 Frequencies of unique alleles in dataset

Alleles	Occurrence	Percentage (%)
C/T	8770	29
A/G	7734	26
A/C/T	539	2
A/C/G	643	2
C/G/T	873	3
C/G	2564	6
A/C	1712	6
A/G/T	588	2
A/T	1159	4
G/T	1798	6
Other		12

ten of the most important alleles and of them, C/T or A/G alleles have 29 and 26% presence in the total dataset. However, ten most influential alleles and their frequencies are illustrated in Table 2.

Besides this, we have analyzed impact of unique alleles in each significance. However, in benign among 10 most influential alleles, C/T and A/G have most frequencies with 30% and 26% respectively. Occurrences of ten most influential alleles in benign are illustrated in Table 3.

Table 3 Occurrences of unique alleles in benign

Alleles	Occurrences	Percentage (%)
C/T	4577	30
A/G	3872	26
A/C/T	237	2
A/C/G	312	2
C/G/T	422	3
C/G	1349	9
-/A	173	1
A/C	899	6
A/G/T	239	2
A/T	570	4
-/G	177	1
-/C	176	1
-/T	181	1
G/T	939	6
Other		6.20

Table 4 Occurrences of unique alleles in malignant

Alleles	Occurrences	Percentage (%)
C/T	2600	29
A/G	2519	28
A/C/T	168	2
A/C/G	185	2
C/G/T	251	3
C/G	791	9
-/A	105	1
A/C	544	6
A/G/T	184	2
A/T	369	4
G/T	555	6
Other		8.10

Similarly, in malignant among 10 most influential alleles, C/T and A/G have most frequencies with 29% and 28% respectively. Occurrences of ten most influential alleles in malignant are illustrated in Table 4.

Apart from these, among unique alleles and highest influencing position, position 26 and alleles C/T have 55% chance the disease to be Benign. Similarly, when position 26 and alleles A/G then there is 54% chance the disease to be Benign. Moreover, when position 51 and alleles A/G then there is 51% chance the disease to be Malignant. However, occurrence of unique alleles and position to each significance is illustrated in Fig. 3.

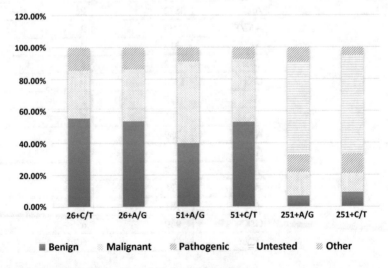

Fig. 3 Occurrence of alleles with position in each significance

4 Methodology

4.1 Convolutional Neural Network

Convolutional neural network (CNN) is an extension of previously referred as artificial neural network (ANN) or Deep Neural Network (DNN). This deep learning architecture allows us to support the main feature of multiple perceptron which is referred as a mechanism required lesser preprocessing of its input matrix compare to the other available architectures. The visual representation of imagery assists convolutional neural network for efficient as well as faster processing of input matrix. Just similar to the other neural network architecture this algorithm is also implemented after being inspired from the actual mechanism of human cortex. Therefore, this deep learning architecture ensures clear demonstration of higher dimensional data. For classification and illustration of various complex as well as convoluted training set this approach is being applied in the field of bioinformatics and disease recognition. Thus, convolutional neural network is the combination of biologically inspired computing formulated [26] by employing various mathematical formulation. From the last decade after invention of this algorithm, revolution has made landfall in the field of machine learning and computer vision and also this approach is being used effectively as well as influentially.

CNN is being focused because of its exceptional ability of learning features automatically from the input dataset. Moreover, less preprocessing mechanism along with dimensionality reduction technique make this approach unique and most potential for complex system learning. The motive aspect of convolutional neural network is the presence of convolutional layer which usually integrate a convolution operation to the input data and passing the output to the following layer. Each neuron of convolutional layer usually deals with the data from its corresponding field otherwise simply ignore it. All the neurons of this layer ensures decreasing robustness as well as complexity of this input data for the confirmation of further efficient processing inside the neuron. The motive task of this layer is to diminish number of parameters so that overall working procedures depicts efficient manipulation of the input data. For accomplishment of this work a low dimension reduction mechanism is employed inside the convolutional neural network code so that the dimension reduction mechanism can be integrated within the CNN framework. CNN consists of an input layer, a hidden layer. Lots of hidden layers and an output layer. Compared to other neural network architecture this convolutional neural network architecture is composed of more hidden layers and all the necessary computation usually takes place within those layers. Here for this specific contribution we have specifically focused on exploration of convolutional neural network for systematic exploration of tuberous sclerosis disease. For characterization of this work we have concentrated on basic infrastructure of convolutional neural network which has been mentioned in Fig. 4.

The input matrix of convolutional neural network always covers total number of input features along with the classification attributes. For this work total 4 hidden

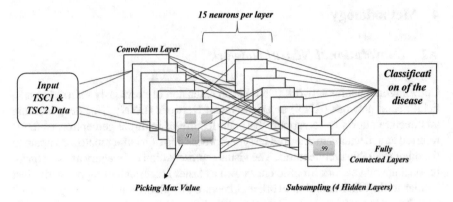

Fig. 4 basic infrastructure of convolutional neural network

layers have been constructed and 15 no of neurons per each layer. In the dataset there are in total eight attributes including classification one. Demonstrating calculation inside the hidden layer is mostly easy and simple, if $M \times M$ has a filter (ω) of $N \times N$ then the size of the output layer of that architecture should be $(M - N + 1) \times (M - N + 1)$. Now for calculating the input of the next layer from the previous layer can be demonstrated as follows:

$$x_{ij}^l = \sum_{k=0}^{N-1} \sum_{l=0}^{N-1} \omega_{kl} y_{(i+k)(j+l)}^{l-1} + bias \tag{1}$$

Now the outcome computed by x_{ij}^l further processed by the following sigmoid activation function to generate output of a specific layer,

$$y_{ij}^l = \sigma\left(x_{ij}^l\right) \tag{2}$$

Using this sigmoid activation function we can finalize exact output of a neuron. Now an error estimation function (E) should be installed by taking derivative ($\frac{\delta E}{\delta y_{ij}^l}$) to calculate the error value from the predicted output and the actual output. The loss function for each neuron is illustrated as follows:

$$\frac{\partial E}{\partial \omega_{ab}} = \sum_{i=0}^{M-N} \sum_{j=0}^{M-N} \frac{\partial E}{\partial x_{ij}^l} \frac{\partial x_{ij}^l}{\partial \omega_{ab}} = \sum_{i=0}^{M-N} \sum_{j=0}^{M-N} \frac{\partial E}{\partial x_{ij}^l} y_{(i+a)(j+b)}^{l-1} \tag{3}$$

Now for stepping towards the optimal solution of the weights and biases value of the derivatives depicted in Eq. 3 need to be further derived. Thus, to calculate value of the derivative term $\frac{\delta E}{\delta y_{ij}^l}$, following equation been considered here,

$$\frac{\partial E}{\partial y_{ij}^{l-1}} = \sum_{a=0}^{N-1}\sum_{b=0}^{N-1} \frac{\partial E}{\partial x_{(i-1)(j-b)}^{l}} \frac{\partial x_{(i-1)(j-b)}^{l}}{\partial y_{ij}^{l-1}} = \sum_{a=0}^{N-1}\sum_{b=0}^{N-1} \frac{\partial E}{\partial x_{(i-1)(j-b)}^{l}} \omega_{ab} \qquad (4)$$

Now by subtracting the derivative value multiplied by learning rate a step towards the optimal weights and biases can be accomplished. Weights and biases update formula can be formed as follows:

$$w_i^l = w_i^l - leqarningrate * \frac{\partial E}{\partial y_{ij}^{l-1}} \qquad (5)$$

$$b_j^l = b_j^l + \Delta b_j^l \qquad (6)$$

Here $\Delta b_j^l = learningrate * \frac{\partial E}{\partial y_{ij}^{l-1}}$.

Repeating every step until convergence of optimal set of weights and then biases can be determined and fit to the neural network for further prediction of new tuples from the outside of training set.

4.2 Particle Swarm Optimization (PSO)

This PSO optimization process is used on such a problem like cyclic order and also it can establish the improved solution than the given solution. Among the all optimizer PSO is an intelligence one. This optimization works such a way to find the parameters which gives the maximum value and it is easy to use as well as it can be implemented dynamically. This algorithm has been developed inspiring from the animal behavior like birds flocking, fish schooling and the one of the most computational fields like the Genetic algorithm.

Figure 5 depicts overall working phases of PSO. PSO can be applied easily without any interference regarding on the problem which need to be optimized. As the Fig. 6, exclaimed that using a built in function rand (.) PSO algorithm is initialized randomly. Then it determines the maximum and minimum value of given function. At the same time, the best-known position are being searched by the each particles of the PSO algorithm as well as best known local particle's position guides the searching process. When the particle gains the best position then automatically it updated the local particle position simultaneously.

The main goal of the PSO algorithm is to chain all the given particle within the optima of a certain dimensional space. The position and the velocity of the individual particles are assigned randomly as the following way. Here, if 'n' number of particles have been considered in a vector form then it can be written as $V = [V_1, V_2, V_3, \ldots V_n]$. First of all the velocity and the position vector of each particle initialized by pseudorandom. It seems that all position vectors are at dihedral angle and that are considered as the phi and psi. Instead of that velocity vector step forward in order to get best known position by changing the angles. This improves

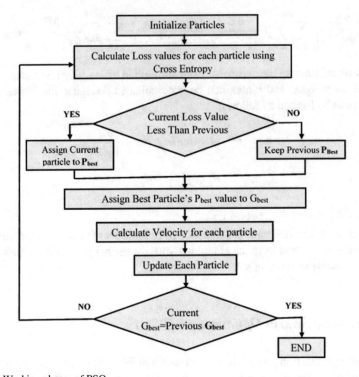

Fig. 5 Working phases of PSO

the flexibility of getting global best-known position. The Eq. (7) shows how to update the velocity.

$$v_i(t+1) = v_i(t) + \left(m_1 \times rand() \times \left(p_i^{best} - p_i(t)\right)\right) + \left(m_2 \times rand() \times \left(p_i^{gbest} - p_i(t)\right)\right)$$

$$(7)$$

From the equation $v_i(t)$ is initial state of the particle vector, $v_i(t+1)$ demonstrate that updated velocity of the each particle, p_i^{gbest} is indicate that global known best position and p_i^{best} is known as kn own best position. Moreover m_1 and m_2 indicates the weights of each particle's global best and personal best position. Besides in order to velocity update, another update for each particle position is required to optima of expected position in every iteration:

$$p_i(t+1) = p_i(t) + v_i(t+1)$$

$$(8)$$

From the equation, the updated position for all the individual particles is $p_i(t+1)$, on the other hand $p_i(t)$ is the previous position before update and the updated velocity of the particle is $v_i(t+1)$. If $f(p_i) > f(p_i(t+1))$, then the best known position will automatically assigned to p_i (*i.e.* $P_i = p_i(t+1)$), g will be the best solution if

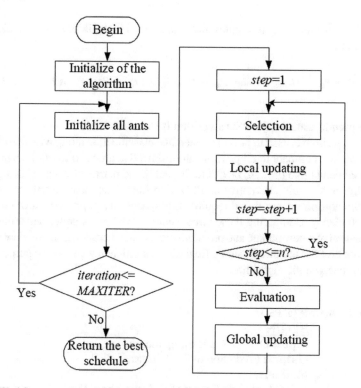

Fig. 6 Workflow procedures of Ant Colony Optimization Approach

$f(p_i) < f(g)$. To represents the PSO algorithm is geometrical form we consider a 2-Dimensional space for experimental purpose. Where there we assume that the particle are moving with their initial positions and velocity. Optimum position: $G_{best} = \min \{p_{best}^t, i\}$, where $i \, \varepsilon \, [1, 2, 3, \dots n]$, $n > 1$. In order to gain optimum coverage the particle will reach G_{best}. The recommended PSO parameters' values are given bellow:

1. From the theory the range of the particles is obtained 25–40, it seem that the range is large enough to get perfect result. Sometimes in order to get better results, need to use more particles.
2. There is a limit of changing the velocity and position for every particle. This criterion is used as stopping criterion.
3. The conceptual limit of weight coefficients m_1 and m_2 are within $[0, 2]$.
4. The stopping criterion depends on the problem to be optimized, but it is terminated when no improvement occurs over some consecutive iterations, then the algorithm stops using:

$$T_{norm} = T_{\max} \div diameter\,(S) \qquad (9)$$

where T_{max} is the maximum radius and $diametr(S)$ is the initial swarm's diameter. In addition,

$$T_{max} = \|P_m - G_{best}\| \text{ With } \|P_m - G_{best}\| \geq |P_i - G_{best}|, \text{ as } i = 1, 2, 3, \ldots n \tag{10}$$

The pseudo code for the PSO algorithm is shown in Table 5.

PSO algorithm is used to provide systematic classification along with that efficient clustering tool compare to all possible algorithms that are used for various bioinformatics elements like protein, DNA etc. In real life experiment it seems that genetic algorithm is absolute alternative of PSO algorithm. The basic deference between these two algorithms are that it requires less spaces and improvement of computational efficiency. Comparing those circumstances PSO gives better performance. A combination of deterministic and probabilistic for each iteration point of view genetic algorithm ad PSO are utterly same. Regarding of efficiency and space purpose PSO is better than genetic algorithm.

Table 5 Pseudocode for PSO

BEGIN
Input: Data D & a learning method
Output: Prediction of TSC disease
Method:
1. **Replace** class Labels of D, loss = limit_INT
2. **Initialize** Particles
3. **For i=0** to num_of_particles:
4.   **Calculate** fitness $= -\sum_{c=1}^{M} y_{o,c} \log(p_{,c})$
5.   **If** (fitness < prev_fitness):
6. $P_{best} = fitness$
7.   **End** if
8. **End** for
9. **For** j=0 to num_of_particles:
10.   **If** (loss > loss (particle)):
11.   Loss $\leftarrow$ loss (particle)
12. $G_{Best} = Particle$
13.   **End** if
14. **End** for
15. **For** k=0 to num_of_particles:
16.   **Calculate** change in velocity
17.   **Update** values of particle
18. **End** for
END

4.3   Ant Colony Optimization (ACO)

Ant colony optimization (ACO) is a probabilistic mechanism inspired from behavior of ants to find an optimal shortest path towards destination. This particular approach is usually applied to reduce time consumption as well as memory requirements by using various statistical probabilistic measures. The main paradigm used here is pheromone based artificial ant behavior. These artificial ants try to locate optimal solution by mathematical computation within a pheromone archive space which represents all possible solutions. More specifically, ant colony optimization technique performs a model-based mechanism to find the optimal solution for any particular problem. For this contribution this ACO approach has been integrated with convolutional neural network for finding the optimal combination of weights and biases so that the neural network model can be trained using those weights and biases. Basic working procedures of ant colony optimization approach is mentioned in Fig. 6.

For this work, basic formulation of ACO has been considered. All the weights and biases have been considered as the pheromone by incorporating them into a solution space referred as repository. Now, after getting loss value for each combination of weights and biases the pheromone table are sorted in descending order. Now the values of weights and biases are updated using the following formulas.

$$\tau(m, s) = (1 - \rho) * \tau(m, s) + \rho * \tau_0 \tag{11}$$

For this equation $\tau(m, s)$ is the values of the weights and biases in each row of the pheromone table. ρ is the evaporation rate and τ_0 is the initial value of the pheromone which is currently being updated. Equation (11) is formulated for updating positive pheromones only whereas both positive as well as negative pheromone should be updated. Therefore, the following formula can be constructed to update both the negative as well as positive values of the pheromones.

$$\tau(m, s) = (1 - \alpha) * \tau(m, s) + \alpha L_{best}^{-1} \tag{12}$$

Here α denotes the best fitness influence degree of the best ant found so far and L_{best} is the loss value of the best ant in the pheromone table. Now further update will be preceded by the following equations,

$$\tau^b(m, s) = \gamma \tau^b(m, s) \tag{13}$$

$$\tau^b(m, s) = \min\{\tau^b(m, s)\gamma^{-1}, 1\} \tag{14}$$

γ Is a constant in the range 0 and 1? $\tau^b(m, s)$ Is a probability measurement to determine whether a pheromone combination of weights and biases should be chosen or not. This is how the weights and biases which have been considered as pheromone are updated in each step and finally optimal combination of weights and biases are found.

4.4 CNN Optimized by PSO

Convolutional neural network nowadays represents better performance for learning
features automatically but still some inconsistencies can be noticed while training
the overall network architecture. Since convolutional neural network architecture
requires much time to be trained and so we have omitted the part of back propagation and employed particle swarm optimization; a nature inspired meta heuristic
approach for updating the weights and biases to make the overall training process
faster than usual. Efficient incorporation of this meta heuristic approach can reveal
better performance compared to back propagation. Overall workflow procedure after
integration of particle swarm optimization is demonstrated in Fig. 7.

Initially, dataset is collected and preprocessed using various preprocessing mechanism so that it can be fed into machine learning approaches. Further layers as well
as neurons are initialized and defined. Also weights and biases are randomly initialized and flattened to form a particle. From those weights and biases by taking
combination a solution space is formulated with 25 particles. After that using loss
function of neural network each particles personal best P_{Best} value is calculated and
updated after comparing with the previous value of personal best. After assignment
of personal best for every particle further best particle G_{Best} is calculated among
all the particles in the swarm. Then value change in velocity is computed as well as
added with the current value of every particle using following formula,

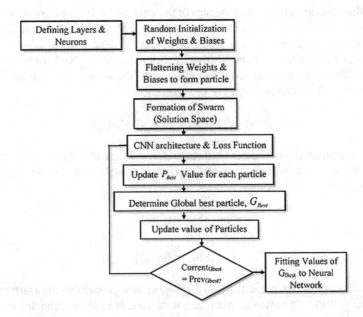

Fig. 7 Overall workflow procedures of neural network optimized by PSO

$$V_i = V_i + C_1 * Rand(0, 1) * (P_{Best} - Cur\_Particle) + C_2 * Rand(0, 1) * (G_{Best} - Cur\_Particle) \tag{15}$$

$$X_i = X_i + V_i \tag{16}$$

Here X_i is the updated new value of particle and C_1, C_2 are the cognitive inertia. Rand () is the random function which returns a value within the range 0–1. The proposed algorithm of convolutional neural network optimized with particle swarm optimization is mentioned in Table 6. Later, a comparative analysis of this approach and neural network optimized with ant colony optimization have been demonstrated in the result analysis section.

For better clarification and understanding the application of PSO on CNN, all the parameters with specific values are referred in Table 7. Those parameters can be assigned to regenerate the findings we have acquired here in this work.

4.5 CNN Optimized by ACO

Ant colony optimization is also one of the Meta heuristic approaches that can be used to find difficult solutions of various optimization problems. In ant colony optimization problem, a set of artificial ants always search for an optimal solution by traversing the whole solution space. All of those artificial ants incrementally traverse the whole solution space for finding optimal number of biases and weights. Overall approach of finding optimal solution is stochastic and always conducted by maintaining a pheromone table inside ant colony optimization approach. This is also one of the probabilistic approaches to solving computational problems which can be employed to solve any specific problem. This approach is one of the members of swarm intelligence family and so for optimizing neural network architecture's weights and biases this specific approach can be employed. Overall workflow procedures of this approach is demonstrated in Fig. 8.

Initially neurons and layers are defined and initialized. Then, formation of pheromone table or archive is conducted by randomly assigning values of weights and biases inside the archive. This phase is followed by calculating error or loss using loss function. The loss function can be demonstrated in Eq. (17)

$$-\sum_{c=1}^{m} y_{o,c} \log(p_{o,c}) \tag{17}$$

After calculating error of each and every row of weights and biases in the pheromone table, all the values are sorted depending on the loss value. Now weights and biases of pheromone table are further updated using the Eqs. (18) and (19).

Table 6 CNN optimized with PSO

<u>**BEGIN**</u>

Input: Data D & a learning method CNN with PSO

Output: Prediction of TSC disease

Method:

1.   Replace class Labels of D with classification labels
2.   **Initialize** No_of_Layers, SizeofNeuron, NeuronPerLayer, No_of_weights, c=0
3.   **Initialize** PSO Module, Swarm of particle
4.   **Derive** Model, M
5.   **Classify,** X_{new} using model, M
6.   **For** i=0 to No_of_layers:
7.   **For** j=0 to No_of_NeuronPerLayer:
8.   **For** k=0 to no_of(weights+biases)_perlayer:
9.   Particle [c++] ← Random values
10.   **End** for
11.   **End** for
12. **End** for
13. **For** l=0 to SwarmSize:
14.   **For** m=0 to no_of(weights+biases)_perlayer:
15.   Index = rand() % **sizeof**(particle)
16.   Swarm[l][m] → Particle [index]
17.   **End** for
18. **End** for
19. **For** I = 0 to MaxIterations:
20.   **For** each particle in **swarm space:**
21.   **Fit weights & biases & calculate loss**
22.   **Calculate** P_{best}
23.   **Update** particle values
24.   **End** for
25. **Find** G_{best}
26. $Model_{new}$ = fit values of G_{best}'s weights and bias into CNN
27. Predict = $Model_{new}(X_{new})$
28. **Return** predict

<u>**END**</u>

$$P_{ij}^k(t) = \frac{[\tau_{ij}(t)]^\alpha \cdot [\eta_{ij}]^\beta}{\sum_{l \epsilon j_i^k} [\tau_{il}(t)]^\alpha \cdot [\eta_{il}]^\beta} \qquad (18)$$

$$\tau_{ij} \leftarrow (1-\rho) \cdot \tau_{ij} + \rho \cdot \sum F(s) \qquad (19)$$

After execution of this phase, further checking is mandatory whether the error of the current solution is less than the threshold value or not. This solution is the 1st

Table 7 Necessary parameters and their values

Parameter	Value	Parameter	Value	Parameter	Value
Iterations	3000	No of layers	5	Cognitive weight (C_1)	2
Test_size	85%	Nodes per layer	10	C_2	2
Learning rate	0.01	Activation function	Tanh	No of weights	10
Loss function	BinaryCrossEntropy	Output activation function	Sigmoid	No of biases	10
Range of weights	$(-1, 1)$	Initial value (Weights)	1	Initial value (Biases)	1

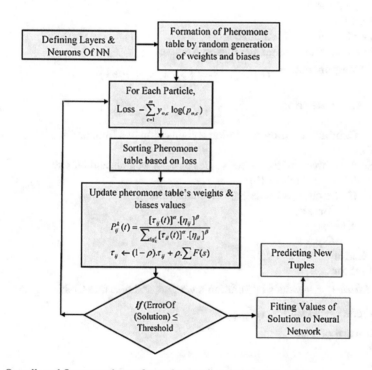

Fig. 8 Overall workflow procedures of neural network optimized with ACO

index of the pheromone table. The Pseudocode for neural network optimized with ant colony optimization is illustrated in Table 8.

For better clarification and understanding the application of ACO on CNN, all the parameters with specific values are referred in Table 9. Those parameters can be assigned to regenerate the findings we have acquired here in this work.

Table 8 CNN optimized with ACO

BEGIN
Input: Data D & a learning method
Output: Prediction of TSC disease
Method:
29. Replace class Labels of D
30. **Initialize** Layers, SizeofNeuron, NeuronPerLayer, No_of_weights
31. **Initialize** Ant Colony Optimization Module, ArchiveSize (Pheromone_Table)
32.   **Derive** Model, M
33.   **Classify**, X_{new} using model, M
34. **For** i=0 to ArchiveSize:
35.   **For** j=0 to No_of(Weights+Biases):
36.   Archive [i] [j] ← Random values
37.   **End** for
38. **End** for
39. **For** I = 0 to MaxIterations:
40. **For** each set of weights & bias in Archive
41.   **Compute** output, $O_{Pr\,edicted} = f(\sum_{i=0}^{m} w_i\,x_i + bias)$
42.   **Compute** error, $E = \dfrac{1}{2m}\sum_{i=0}^{m}(O_{actual} - O_{predicted})^2$
43.   **Compute & update** solution weights & biases from archive table
44. **End** for
45. **Sort** Archive based on error values of each weights and bias set
46.   Solution ← Archive [0]
47.   **If** (ErrorOf (Solution) ≤ Threshold value):
48.   **Break;**
49.   **Else:**
50.   _**Repeat;**_
51.   **End** if
52. **End** for
53. $Model_{new}$ = values of Solution's weights and bias into CNN
54. Predict = $Model_{new}(X_{new})$
55. **Return** predict
END

5   Result Analysis

Different performance measurement has been utilized including accuracy, timing and memory requirements for the cost for iteration graph etc. In addition, cost per iteration graph has also been presented for comparative study.

Table 9 Necessary parameters and their values

Parameter	Value
MaxIterations	10000
Test_size	80%
Evaporation rate	0.85
Loss function	BinaryCrossEntropy
Single pheromone	No of weights and biases in NN
No of pheromone	20

Accuracy comparison is the most common and widely utilized performance measurement metric for machine learning approaches. Accuracy is by using the following formula:

$$Accuracy = \frac{(TP + TN)}{(TP + TN + FP + FN)}$$

where, TP, TN, FP, FN are True Positive, True Negative, False Positive and False Negative respectively. Comparisons of the accuracies of all the approaches are mentioned in Table 10.

CNN with PSO achieves the highest accuracy among all the applied algorithms. While CNN with ACO performed better than CNN but not better that some of the methods, although the differences are too small and can be ignored. From the comparison, it can be mentioned that not all the algorithms are suitable for such experiments.

While comparing the accuracies it is also required to check the time and memory requirements for training the samples of the proposed algorithms to understand their effectiveness as well as applicability. Table 11 illustrates time and memory usage of only CNN optimized with PSO and ACO.

Table 10 Comparison of accuracy of the algorithms

Algorithm	Accuracy (%)
Support vector machine	81.13
Naive bayes	80.13
K nearest centroid	78.79
Logistic regression	80.13
Logistic regression CV	79.79
K nearest neighbor	78.79
Decision tree	80.79
Random forest	78.21
CNN	76.82
CNN with PSO	**83.47**
CNN with ACO	**80.04**

Table 11 Performance comparison of the proposed approach

Name of the approach	Memory usage (MB)	Time required for training (sec)	Accuracy (%)
CNN	**339.130**	163.21	76.82
CNN optimized with PSO	514.84	**142.81**	**83.47**
CNN optimized with ACO	475.34	**148.21**	**80.04**

Table 12 Classification accuracy by MYRA without cross validation

Dataset no	Accuracy (%)	Dataset no	Accuracy (%)
1.	60.91	10.	55.30
2.	49.97	11.	55.77
3.	54.17	12.	43.10
4.	44.16	13.	71.05
5.	62.21	14.	48.43
6.	64.91	15.	52.90
7.	42.76	16.	44.10
8.	73.45	17.	62.71
9.	64.18	18.	57.58
Average accuracy			**55.98**

However, memory requirements of both of our proposed algorithms are higher than CNN but their training times are less than that of CNN and of course classification accuracies are much better.

We also use MYRA [18] an open source java-based tool to check the performance of ACO. However, the tool can only take 1.5 k instances of data at a time. As our dataset is contains a huge number of instances, we divided it into 24 sets and apply MYRA to check the performance. The classification on those smaller datasets are conducted using with cross validation and without cross validation. The classification results shown in Tables 12 and 13.

5.1 Association Rule Construction

For generating association rule, we have implemented both the Decision Tree and Apriori. We have also carried out manual generation of association rules from the dataset because we are confident of the results of our implemented approach. The rules are represented as a set of antecedents and consequents in IF-THEN format.

Table 13 Classification accuracy depicted by MYRA with cross validation

Dataset no	Accuracy (%)	Dataset no	Accuracy (%)
1	55.45	2	55.13
3	55.57	4	53.70
5	54.09	6	52.77
7	55.61	8	55.08
9	54.84	10	54.75
11	54.53	12	54.63
13	53.22	14	54.05
15	55.81	16	54.69
17	53.09	18	55.15
19	54.71	20	58.17
21	55.24	22	54.31
23	54.14	24	54.13
Average accuracy			**54.70**

The association rules produced after implementing Apriori algorithm on our dataset are described in Table 14.

Association rules are also generated by applying decision trees by traversing the trees from root to the leaf. Association rules generated by decision tree are listed in Table 15. These generated rules are almost similar to those by Apriori algorithm.

Table 14 Association rules generated from apriori algorithm

R1: $\left(\begin{array}{c}\text{Length} = 1001 \wedge \text{Position} = 501 \wedge \text{Significance} = \text{Benign} \\ \wedge \text{ Alleles} = \text{T/C} \wedge \text{Allel}_{\text{Origin}} = \text{C/T}\end{array}\right) \rightarrow$
(Sequence = CTCTGAAGGTCCAGGTCCAAAGAGTTTCTGCAAAGTGTAT ...)

R2: $\left(\begin{array}{c}\text{Length} = 801 \wedge \text{Position} = 401 \wedge \text{Significance} = \text{Pathogenic} \\ \wedge \text{Alleles} = \text{A/G} \wedge \text{Allel}_{\text{Origin}} = \text{G/A}\end{array}\right) \rightarrow$
(Sequence = GATAGTAGGTGGCAATCTTAAGTGTGAACACTTCC ...)

R3: $\left(\begin{array}{c}\text{Position} = 251 \wedge \text{Significance} = \text{Benign} \wedge \text{Alleles} = \text{A/G} \wedge \\ \text{Allel}_{\text{Origin}} = \text{G/A}\end{array}\right) \rightarrow$
(Sequence = TACATTTTTAGGTTGGGCACCTTCA)

R4: (Length = 51 ∧ Position = 26 ∧ Significance = Pathogenic ∧ Alleles = T/G)
→ (Allel_Origin = G/T ∧ Significance = Pathogenic)

R5: (*Length* = 1001 ∧ *Position* = 501 ∧ *Significance* = Benign ∧ *Alleles*
= *T/G / C/A* ∧ *Allel_Origin* = *A/C / G/T*) → (*Sequence*
= ACGGAACCAGGAAACTAGACTGTATTGGGTTTTAAGCTTTCCTTT ...)

R6: (*Length* = 26 ∧ *Position* = 51 ∧ *Alleles* = *T/C*) → (*Significance*
= *Malignant* ∧ *Allel_Origin* = *C/T*)

R7: (Position = 251 ∧ Significance = Pathogenic ∧ Alleles = A/G) → (Sequence
= CAGCTGTCATTAGCAGGACCCGTAC)

R8: (Length = 1001 ∧ Position = 501 ∧ Significance = Benign ∧ Alleles
= A/C / G/T ∧ Allel_Origin = T/G / C/A) → (Sequence
= ACGGAACCAGGAAACTAGACTGTATTGGGTTTTAAGCTTTCCTTTG ...)

Table 15 Association rules by decision tree

R1: (Position = 26 ∧ Length = 51 ∧ Alleles = T/C) → (Significance
= Malignant ∧ Allel_Origin = C/T)
R2: (Length = 1001 ∧ Position = 501 ∧ Significance = Benign ∧ Alleles
= C/T ∧ Allel_Origin = T/C) → (Sequence
= CTCTGAAGGTCCAAAGAGTTTCTGCAAAGTGTATGTG ...)
R3: (Position = 251 ∧ Significance = Benign ∧ Alleles = A/G ∧ Allel_Origin = G/A)
→ (Sequence = TACATTTTTAGGTTGGGCACCTTCA)
R4: (Length = 51 ∧ Position = 26 ∧ Significance = Pathogenic ∧ Alleles = T/C)
→ (Allel_Origin = C/T ∧ Significance = Pathogenic)
R5: (Length = 801 ∧ Significance = Other ∧ Alleles = G/A ∧ Allel_Origin = A/G)
→ (Position = 401 ∧ Sequence = GATAGTAGGTGGCAATCTTAAG ...)
R6: (Length = 26 ∧ Position = 51 ∧ Alleles = T/C) → (Significance
= Malignant ∧ Allel_Origin = C/T)
R7: (Position = 251 ∧ Significance = Pathogenic ∧ Alleles = A/G) → (Sequence
= CAGCTGTCATTAGCAGGACCCGTAC)
R8: (Length = 51 ∧ Position = 26 ∧ Alleles = T/C) → (Allel_Origin = C/T ∧ Significance
= Pathogenic)

The association rules show some hidden relationship among the features that were not evident from the outset of the dataset.

6 Conclusion

Tuberous Sclerosis is a rare genetic multisystem disorder disease which in the long run results in Epilepsy and Seizures. This disease can cause tumors in various parts of the human body. Lots of research have been already conducted to find significant information to make a cure for this disease. In spite of all the invention, no effective automated mechanism has been found yet to detect or identify any pattern for finding cure which is mandatory in this era of big data. Manual procedure is not feasible any more.

In order to mitigate the challenge, a hybrid mechanism is proposed in this chapter where two nature inspired algorithms [27], namely PSO and ACO are combined independently with CNN. Experiments have been conducted where each of the algorithms are applied independently and the performance of the each of the techniques are compared with the hybrid algorithms. The obtained results are promising as each hybrid algorithm performs better than that of the originals. Several other commonly used data mining techniques are also applied to the dataset to perform a comparative analysis of the proposed algorithms. The experimental results reveal that the proposed algorithms also outperform most of the cases.

The experiments reveal the effectiveness of applying nature inspired computing techniques in data sciences. Such result inspires the wider adoption of these

approaches in data science applications. When PSO and ACO are applied independently, their performances are average. While combining CNN with these optimization techniques, taking the best of the both approaches the results are promising.
Relationship among the attributes of a dataset also reveals several critical information regarding the disease. This chapter generates various rules illustrating the
relationship among the attributes by applying apriori algorithm. The rules unveil
significant hidden relationships among the attributed towards classifying the disease
and they can play a key role while detecting and predicting the disease.

While it is evident that adopting nature inspired computation techniques in data
science application significantly improves the computational accuracies, we are still
looking forward to enhancing the experiment to improve the results even further and
get more detailed insight about such disease. Hence, we are planning to applying
map-reduce technique to the hybrid algorithms to fit these techniques in Big Data
framework. Further to this it is also in our future plan to apply other evolutionary
approaches to analyzing and predicting of such deadly disease.

References

1. Sahin, M., et al. 2016. Advances and future directions for tuberous sclerosis complex research: Recommendations from the 2015 strategic planning conference. *Pediatric Neurology* 60: 1–12.
2. Humphrey, A., J.N.P. Higgins, J.R.W. Yates, and P.F. Bolton. 2004. Monozygotic twins with tuberous sclerosis discordant for the severity of developmental deficits. *Neurology* 62 (5): 8–795.
3. Nelson, S.L., and B.M. Wild. 2018. Tuberous sclerosis complex: Early diagnosis in infants. *Pediatric Neurology Briefs* 32: 12.
4. Słowińska, M., et al. 2018. Early diagnosis of tuberous sclerosis complex: A race against time. How to make the diagnosis before seizures?. *Orphanet Journal of Rare Diseases* 13 (1): 25.
5. Sancak, O. 2005. *Tuberous Sclerosis Complex : Mutations, functions and phenotypes.* PrintPartners Ipskamp.
6. Roach, E.S., F.J. DiMario, R.S. Kandt, and H. Northrup. 1999. Tuberous sclerosis consensus conference: Recommendations for diagnostic evaluation. *Journal of Child Neurology* 14 (6): 401–407.
7. Lindhout, D., and M. Goedbloed. Analysis of TSC2 stop codon variants found in tuberous sclerosis patients. *European Journal of Human Genetics.*
8. M. European Society of Human Genetics, et al. 1993. *European Journal of Human Genetics* 17 (3). Karger.
9. Niall, H.D. 1971. Automated sequence analysis of proteins and peptides. *Journal of Agriculture and Food Chemistry* 19 (4): 638–644.
10. Niall, H.D. 1973. Automated edman degradation: The protein sequenator. *Methods in Enzymology* 27: 942–1010.
11. Chao-Yuh, Y., and F.I. Sepulveda. 1985. Separation of phenylthiocarbamyl amino acids by high-performance liquid chromatography on spherisorb octadecylsilane columns. *Journal of Chromatography A* 346: 413–416.
12. Venable, J.D., M.-Q. Dong, J. Wohlschlegel, A. Dillin, and J.R. Yates. 2004. Automated approach for quantitative analysis of complex peptide mixtures from tandem mass spectra. *Nature Methods* 1 (1): 39–45.
13. Zou, An-Min, Jiarui Ding, Jin-Hong Shi, and Fang-Xiang Wu. 2008. Charge state determination of peptide tandem mass spectra using support vector machine (SVM). In *2008 8th IEEE international conference on bioinformatics and bioengineering*, 1–6.

14. Ess, K.C. 2009. Tuberous sclerosis complex: Everything old is new again. *Journal of Neurodevelopmental Disorders* 1 (2): 9–141.
15. Dragoumi, P., F. O'Callaghan, and D.I. Zafeiriou. 2018. Diagnosis of tuberous sclerosis complex in the fetus. *European Journal of Paediatric Neurology* 22 (6): 1027–1034.
16. Leal, A.J.R., A.I. Dias, J.P. Vieira, A. Moreira, L. Távora, and E. Calado. 2008. Analysis of the dynamics and origin of epileptic activity in patients with tuberous sclerosis evaluated for surgery of epilepsy. *Clinical Neurophysiology* 119 (4): 853–861.
17. N. D. Human Genome Variation Society, et al. 1992. *Human Mutation* 26 (4). Wiley-Liss, Inc.
18. Otero, Fernando E.B. MYRA: A Java ant colony optimization framework for classification algorithms. In *Proceedings of the genetic and evolutionary computation conference companion* (GECCO '17), 1247–1254. New York, NY, USA: ACM.
19. Kennedy, J., and R. Eberhart. 1995. Particle swarm optimization. In *Proceedings of ICNN'95-international conference on neural networks*, vol. 4, 1942–1948. Perth, WA, Australia.
20. Dorigo, M., M. Birattari, and T. Stutzle. 2006. Ant colony optimization. *IEEE Computational Intelligence Magazine* 1 (4): 28–39.
21. Das, H., B. Naik, and H.S. Behera. 2018. Classification of diabetes mellitus disease (DMD): A data mining (DM) approach. In *Progress in computing, analytics and networking*, 539–549. Singapore: Springer.
22. Sahani, R., C. Rout, J.C. Badajena, A.K. Jena, and H. Das. 2018. Classification of intrusion detection using data mining techniques. In *Progress in computing, analytics and networking*, 753–764. Singapore: Springer.
23. Das, H., A.K. Jena, J. Nayak, B. Naik, and H.S. Behera. 2015. A novel PSO based back propagation learning-MLP (PSO-BP-MLP) for classification. In *Computational intelligence in data mining*, vol. 2, 461–471. New Delhi: Springer.
24. Pradhan, C., H. Das, B. Naik, and N. Dey. 2018. *Handbook of research on information security in biomedical signal processing*, 1–414. Hershey, PA: IGI Global.
25. Sahoo, A.K., S. Mallik, C. Pradhan, B.S.P. Mishra, R.K. Barik, and H. Das. 2019. Intelligence-based health recommendation system using big data analytics. In *Big data analytics for intelligent healthcare management*, 227–246. Academic Press.
26. Pattnaik, P.K., S.S. Rautaray, H. Das, and J. Nayak. 2017. Progress in computing, analytics and networking. In *Proceedings of ICCAN*, vol. 710.
27. Nayak, J., B. Naik, A.K. Jena, R.K. Barik, and H. Das. 2018. Nature inspired optimizations in cloud computing: applications and challenges. In *Cloud computing for optimization: foundations, applications, and challenges*, 1–26. Cham: Springer.
28. Koohi, I., and V.Z. Groza. 2014. Optimizing particle swarm optimization algorithm. In *2014 IEEE 27th Canadian conference on electrical and computer engineering (CCECE)*, 1–5. Toronto, ON. https://doi.org/10.1109/ccece.2014.6901057.

Mid-Term Home Health Care Planning Problem with Flexible Departing Way for Caregivers

Wenheng Liu, Mahjoub Dridi, Hongying Fei and Amir Hajjam El Hassani

Abstract Home Health Care (HHC) centers aim to deliver health care service to patients at their domiciles to help them recover in a convenient environment. Staffs planning in HHC centers have always been a challenging task because this issue not only just dispatches suitable caregivers to serve patients with considering many peculiar constraints, such as, time window of patients, qualification and mandatory break of caregivers, but also seeks optimal solution to achieve its objective, which is often taken to equivalent to minimize total operational cost, or to maximize satisfaction of patients or caregivers. This chapter formulates a mixed integer programming model for mid-term HHC planning problem, which aims at minimizing the total operational cost of HHC centers. We define the real-life constraints as follows: patients need to be visited once or for several times during the planning horizon by capable caregivers; patients must be served in their time window; each patient has specific preferences to caregivers for some personal reasons (e.g. gender); caregivers work in their contract working time with no more than daily maximum working time; a lunch break happens only if caregivers start to work before the lunch start time and finish working after the lunch end time. Specially, in real life, caregivers can use their own cars or rent cars from HHC center to complete their service tasks, and thus, this chapter firstly concerns a flexible departing way for caregivers, which means that each caregiver can either start working from their domiciles or from HHC center according to the transportation mode chosen though they must end their work at HHC center. We call this way of providing service by caregivers as Center and Domicile to

---

W. Liu · M. Dridi · A. H. E. Hassani
Nanomedicine Lab, University Bourgogne Franche-Comté, UTBM, 90000 Belfort, France
e-mail: wenheng.liu@utbm.fr

M. Dridi
e-mail: mahjoub.dridi@utbm.fr

A. H. E. Hassani
e-mail: amir.hajjam@utbm.fr

H. Fei (✉)
School of Management, Shanghai University, 99 ShangDa Road, Baoshan District, Shanghai 200444, China
e-mail: Feihy@shu.edu.cn

© Springer Nature Switzerland AG 2020
M. Rout et al. (eds.), *Nature Inspired Computing for Data Science*,
Studies in Computational Intelligence 871,
https://doi.org/10.1007/978-3-030-33820-6_2

Center (CDC). In addition, in order to discuss the relationship between caregivers' geographical areas and optimal results, we put forward the other two scenarios, (1) each caregiver must only depart from and return to the HHC center, and this case is named as Center to Center (CC); (2) each caregiver departs from their own domicile and returns to HHC center, this scenario is named as Domicile to Center (DC). As the departing ways for caregivers keep the same in these two scenarios, they can be called fixed departing ways. All these models are solved by a commercial programming solver Gurobi through the two group modified classic Periodic Vehicle Routing Problem with Time Windows (PVRPTW) benchmark instances and the other two group instances generated randomly. In total, 21 instances with up to 12 caregivers, 20 patients and 47 demands during the planning horizon. Experimental results show that the model CDC is the best model as it can find optimal solution for 90% instances, and solve the problem with least computational time for 40% instances. Model CC gets the worst results with no optimal solution can be obtained for all instances. This work will help HHC centers planners make proper decision through managing the depart way of caregivers that satisfy the real life constraints, minimize the total operational cost as well as find the optimal solution efficiency.

1 Introduction

The main issue of HHC planning problem is to dispatch skilled caregivers to serve elder and disabled people at their homes; in many countries, a planning in HHC centers is completed with a mid-term planning horizon to guide a multi-day task. However, over the past decades, due to the aging of the population, dramatic changes in the social living situation and the development of innovative technologies in healthcare area, for example, Big Data analytics [1], there is no doubt that HHC services are in high demand in the future. As a result, efficient planning is becoming more difficult and time consuming. On the other hand, HHC has a great market prospects as their costs are lower than hospital health care, and also rising financial investment in its health care area. The health care industry has become one of the largest sectors of the economy mainly in North America and Europe [2]. The United States is the largest share in the targeted market of HHC in 2011; approximately 4,742,500 patients received services from HHC centers. And in 2012, 143, 600 full-time equivalents were employed by 12,200 home health agencies registered [3], and the HHC market was assessed at 77.8 billion dollars [4]. Europe got the second largest share in which around 1–5% of total public health budget was spent on HHC service [5], in Germany, for example, the number of people dependent on HHC services has already reached 1.6 million in 2012 and is still increasing [6]. All these factors contribute to our choosing HHC as a valuable field for research.

Generally, in order to schedule multi-day HHC planning, senior nurses collect all patients' demands ahead, and assign caregivers to serve these demands according to peculiar constraints. For patients, they need to be visited by capable and preferable caregivers within predefined time window, and for caregivers, contract working time; daily maximum working time and lunch break have to be considered. Finally, caregivers receive their work schedules for the coming several days. The objective of HHC planning can be described from several aspects (e.g. minimize total cost, and maximize satisfaction of patients or caregivers). It depends on HHC centers themselves to decide which objective is the most important for them. In real life, it is almost impossible task to manually draw up an efficient planning considering so many constraints and various demands of patients.

Additionally, in most studies, caregivers can only start their visit from HHC center to serve the demands then back to HHC center. But in practice, this is not convenient for caregivers because they have to go to HHC center first and then start to work. Obviously, caregivers will be more satisfied if they can begin the visit from their homes. For HHC centers, if caregivers' geographical areas are closer to domiciles of patients than HHC center, it would incur lower total costs for caregivers to depart from their homes than HHC center; therefore, with the actual operations taken into account, a novel departing way is needed to satisfy caregivers.

The main contribution of this chapter is to presents a novel mathematical model for the mid-term HHC planning problem with considering multiple complexes real-life constraints and to propose a flexible departing way CDC model for caregivers which have never been considered before. Besides, the other two fixed departing ways CC and DC for caregivers have been also developed to be compared with the flexible way in order to discuss the effect of caregivers' geographical area on the objective function. All models are validated through adapted classic PVRPTW benchmark instances by Gurobi a well-known commercial solver.

The rest of the chapter is organized as follows: in Sect. 2, an overview of the current literature concerning this problem is given. Section 3 describes the detail problem and the mathematical model. Section 4 illustrates the experimental results. The conclusions and perspectives are given in the Sect. 5.

2 Literature Review

Considering the planning horizon, HHC planning problem usually consists of single-day planning problem and mid-term planning problem. In this chapter, a literature review is proved from these two perspectives.

2.1 Single-Day Planning Problem

According to the literatures, many studies dealing with the daily planning problem of HHC services can be observed. To the best of our knowledge, Begur et al. [7], Chen and Rich [8] were among those who first addressed this problem. With the development of methodological process and computer technologies, researches have increasingly focus on more complicated objectives and deal with various real-life constraints. Most literatures considered that each caregiver must start and end their visit in a single HHC center, while multiple depots of departure and arrival of caregivers has been less studied. Bektas [9] first reviewed the multiple traveling salesman problems. Akjiratikarl et al. [10] first proposed a case about multi-depots in which multiple type of HHC service exists. Kergosien et al. [11] presented a study that covered both single and multiple depots by dissociating the starting and ending locations for each caregiver. Furthermore, from a practical point of view, a caregiver also can depart from and return to their home. Trautsamwieser et al. [12] introduced three types of routes according to the starting and ending locations: routes start and end at the hospital, routes start and end at home, and routes start and end at home but the route duration between the caregiver's home and the patient are not considered in the working time. Considering the qualification index, Eveborn et al. [13] first presented a decision support system with considering qualifications of caregivers, the qualifications level is hierarchical, and a caregiver at higher level can cover the services requesting for lower level qualification while the opposite is obviously not true, more details about this point can be found in [14–16]. In addition, other authors also considered preference of patients as caregivers might be rejected by patients for language incompatibility or personal reasons [12, 17]. Moreover, Braekers et al. [18] presented a bi-objective home care scheduling problem in which preference of patients is one of the objectives, which represents the service level of a HHC center, in this study, patients' preference can be treated as a soft constraint by proposing the penalty if it is not satisfied. Another case is that part of the patients must be visit with a preference ratio [19]. The break for caregivers is also a real-life constraint in HHC centers. In most studies, authors mentioned mandatory breaks by setting a maximum working time; caregivers must take a break if their working time exceeds maximum working time [20, 21]. In recent years, some researchers considered the break from another point of viewpoint. Xiao et al. [19] considered the flexible lunch break of caregivers innovatively due to the fact that there is an interval of lunch time, if the caregiver is assigned to work before interval and meanwhile he/she finishes work after interval, the caregiver should take a lunch break during this interval.

2.2 Mid-Term Planning Problem

Over the past decade, researchers have also dealt with mid-term planning of HHC services. Chen et al. [22] proposed a Markov decision process model for the multi-period technician scheduling problem with experience-based service times and stochastic customers, and they solved the model by applying an approximate dynamic programming-based solution approach. Moussavi et al. [23] presented a sequencing generalized assignment model for multiple-day HHC planning problem, and a mat-heuristic approach based on the decomposition of the formulation is introduced to simplify the mathematical model and reduce the computational time.

Except for the restrictions mentioned in single-day planning problem, there are some special constraints in mid-term planning such as visits frequency, contract working time and workload balance of caregivers. Visits frequency means that patients need to be served a few times during each planning horizon. Some researchers assumed a regular frequency of visits, which refers to the principle that a set of services for a given patient should occur at the same day each week. Bennett and Erera [24] defined a Home Health Nurse Routing and Scheduling (HHNRS) problem with time consistency constraints, in which patients must be served on the same day and time of weekly visits, and a rolling horizon myopic planning approach with using a new capacity-based insertion heuristic is developed to obtain the solutions. Duque et al. [25] developed a decision-support system to support the home care service planning for a "social profit" organization, in this system, the visits are evenly distributed over the week, the patients are visited during the same time slots every week, and a flexible two-phase algorithm is employed to tackle the problem. Liu et al. [26] presented a periodic home health care logistics in which a constraint minimum number visits in the planning horizon for each patient is considered. Some articles also characterized patients in a set of days and a visit can only scheduled on a day from this set [27, 28].

In HHC centers, caregivers have their contracted working time, during which they can provide care services to patients, and overtime is either not allowed at all [29, 30], or allowed with caregivers be paid with additional compensation for overtime hours [6]. The caregivers cannot also exceed maximum daily time per day, and keep minimum rest time between two consecutive working days [31]. In addition, in order to meet caregivers' working time constraints, workload balancing also should be also guaranteed to keep caregivers motivated and reduce burnout within HHC centers [32]. More details about HHC planning problem can be seen in [33, 34].

To sum up, although more and more researchers are interested in developing planning strategies for HHC centers, from both short-term and mid-term views, most of them assumed the HHC center to be the only departing point of the caregivers though it will be more efficient for the caregivers to depart from their own domiciles when they use their private cars as working vehicles. In this study, the constraint that all paths start and end at the same depot will be relaxed with a purpose not only to improve the satisfaction of the staff but also to reduce the operational cost.

3 Problem Descriptions and Mathematical Model

3.1 Problem Descriptions

HHC is a growing industry in the medical services business, and these care services are provided by developing a planning to dispatch caregivers to patients' locations, usually, this planning is made with a mid-term planning horizon to guide a multi-day work for HHC center. A typical HHC planning is completed in the end of a week. Next week, related caregivers complete their visits based on this planning. Specially, in some HHC centers, caregivers can depart either from their homes or from HHC center, though all of them must finally go back to the HHC center to summarize their daily works. The flexibility of departing way for caregivers depends on whether they use their private cars or rental cars as the working vehicles. If the caregivers use their private cars, they may depart from their homes, and if the caregivers need to rent a car for consecutive days, they start the visits from their home so that they can keep the rental cars at home for the work of the coming day; only if they rent a car one day, but the day before they use their own cars, caregivers start working from the HHC center because they need to get the rental car first, and hence there is no doubt that in this way of management, HHC center will be able to improve the satisfactions of the caregivers.

In general, the caregivers send a request to HHC center about which day they need to rent a car and which day they can use their own car, according to this request, the center rents the cars according to the requirements of the caregivers for the following week. As shown in Table 1 is the car rental request information of five caregivers during one planning period, i.e. a week.

In general, HHC centers provide several services for patients, such as cleaning, rehabilitation training, and Table 2 shows a part of data that presents the services needed by patients during the planning horizon. Patients can be treated once or several times in the planning horizon within their time window, each patient is visited at most once a day, further, it is noted that patients prefer some caregivers for some personal reasons (e.g. gender), and the basic information about patients can be seen in Table 3.

Only the caregivers qualified no lower than the level required by the treatment of a patient can be assigned to serve that patient. Each caregiver has a working time

Table 1 Car rental demands required by a set of caregivers

Caregivers	Day1	Day2	Day3	Day4	Day5
K1	Rental[a]		Rental	Rental	
K2	–	Rental	Rental		
K3	Rental			Rental	
K4	Rental	Rental		Rental	Rental
K5			Rental	Rental	Rental

[a]Request of renting a car

Table 2 An example of services required by patients during the planning horizon

Patients	Day1	Day2	Day3	Day4	Day5
P1	Service1		Service2	Service1	
P2	Service3	Service2			
P3		Service1		Service4	
P4	Service3		Service1		Service2
P5			Service4		
P6		Service1	Service3		

Table 3 An example of patients' information

Patients	Time window	Preferred caregiver
P1	10:00–12:00	Caregiver 1, 4
P2	9:30–13:00	Caregiver 2, 4
P3	14:00–18:00	Caregiver 1, 3, 4
P4	9:00–12:00	Caregiver 1
P5	11:00–16:30	Caregiver 1, 2
P6	10:30–14:00	Caregiver 2, 3

Table 4 An example of caregivers' information

Caregivers	Contract working window	Qualified services
K1	8:30–18:30	Service 1, 2, 3, 4
K2	9:00–19:00	Service 1, 2
K3	10:00–20:00	Service 3, 4
K4	8:00–18:00	Service 1, 2, 3
K5	8:30–18:30	Service 2, 3

on each day, and all the tasks should be assigned during this interval. In addition, according to the contract, the maximum working time per day, which is set forth in the contract, a lunch break has to be considered as constraints. The information about caregivers is presented in Table 4.

The objective of this problem is to minimize the total operational cost of HHC service during the planning horizon by taking into account the penalty for unscheduled visits, wages of caregivers, fuel cost and charges for operational vehicles. Assumptions in this study are as follows:

1. Departure points of the caregivers can either be their homes (hereinafter referred to as "locations of caregivers") when they use their own car as working vehicle or the HHC center when they use a rental car while each journey ends up at the HHC center because all caregivers must returns to the HHC center to report the daily work summary after finishing their daily schedule. The route shown in Figs. 1 and 2 correspond to the former and the latter respectively, where three

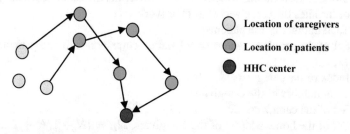

○ **Location of caregivers**

◐ **Location of patients**

● **HHC center**

Fig. 1 Caregivers depart from their homes and return to HHC center

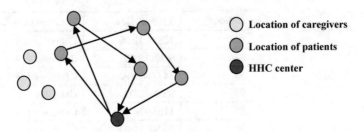

Fig. 2 Caregivers start and end their work at HHC center

kinds of nodes are involved: locations of patients, locations of caregivers and a HHC center;

2. Traffic condition is not considered thus the travel time of the caregivers between two locations is estimated according to their distance;
3. Costs of the oil consumed per distance unit by a rented car and caregiver's car are same, also charges for a rented car and caregiver's car are the same;
4. A penalty is considered if a patient cannot be served as scheduled;
5. Caregivers work as a full-time employee and refuse to work overtime, if there are no missions for serving patients on some days, they will work at HHC center. Besides, each caregiver has the same basic salary, and if they be assigned to serve patients, he/she will get additional income according to their working time;
6. A lunch break should be considered for the caregivers in case that their working time covers the lunchtime. The lunch break is payed the same as normal working time;
7. Caregivers qualified with higher level can serve the patients asking for lower level service, but not vice versa.

3.2 Mathematical Model

Parameters

i, j	Index of the locations which correspond to either the addresses of the patients or the sites the caregivers start their work;
N	Total number of the patients;
Ω_P	Set of the locations where each location corresponds to one patient, $\Omega_P = \{1, ...i, ..., N\}$;
k	Index of the caregivers;
K	Total number of the caregivers;
Ω_k	Set of the caregivers, $\Omega_k = \{1, ...k, ..., K\}$;
Ω_D	Set of the home address of the caregivers, $\Omega_D = \{N + 1, N + 2...N + k, ..., N + K\}$;

$\{0,C\}$	Set of the locations where 0 is the origin of the routes if caregivers start their work from HHC center, C is the destination of all routes. In this chapter, 0 and C all represent the HHC center.
t	Index of the day in planning horizon;
T	Total number of days covered by the planning period;
Ω_T	Set of the days covered by the planning period;
q	Index of the qualification level of the caregivers' professional skill.
Q	Total number of the qualification levels;
Ω_Q	Set of the possible qualification levels. In general, the qualification levels are hierarchical, 1 is the lowest level while Q is the highest level, $\Omega_Q = \{1, 2..., Q\}$;
bP^{kq}	= 1 if the qualification level of caregiver k is no lower than q; 0 otherwise.
WL^{kt}	The earliest working time of caregiver k on day t;
WU^{kt}	The latest working time of caregiver k on day t;
PL_i^t	Lower bound of patient i's time window on day t, if there is no demand of patient i on day t, it equals 0;
PU_i^t	Upper bound of patient i's time window on day t, if there is no demand of patient i on day t, it equals 0;
LB	Universal beginning time of the lunch break for the caregivers;
LE	Universal ending time of the lunch break for the caregivers;
s_i^t	Service duration of patient i on day t;
tt_{ij}	Travel time from location i to location j;
d_{ij}	Travel distance from location i to location j;
n_i^{qt}	Number of caregivers with at least qualification level q that patient i needs on day t;
C^P	Penalty cost if any patient is unable to be served;
L^{break}	Lunch break duration which is fixed for each caregiver k on each working day t;
L	Maximum working time duration per day for each caregiver;
C^W	Wage of each caregiver per time unit;
C^O	Cost of the oil consumed by one car per distance unit;
C^V	Daily charge of each car;
M	A large positive constant to keep the model linearly;

Decision Variables

p_i^k	Binary variable, equals 1 if caregiver k is preferred by patient i, 0 otherwise;
w_i^k	Binary variable, equals 1 if caregiver k stars his/her work from their home's location i, 0 otherwise;
z^{kt}	Equals 1 if caregiver k rents a car from HHC center service on day t, 0 otherwise;
x_{ij}^{kt}	Binary variable, equals 1 if location j is visited after location i by caregiver k on day t, 0 otherwise;
d_i^t	Binary variable, equals 1 if the patient at location i demands a service on day t;
u_i^t	Binary variable, equals 1 if the patient at location i is unable to be served on day t, 0 otherwise;

h_i Artificial variable for constructing the sub-tour elimination constraint;

tL^{kt} Time point that caregiver k starts to take a lunch break on day t;

tB^{kt} Starting time point of caregiver k at their home or HHC center on day t;

tE^{kt} Ending time point of caregiver k at the HHC center on day t;

ts_i^{kt} Time point when caregiver k at location i on day t;

$v1^{kt}$ Binary decision variable equals 0 if caregiver k starts working earlier than beginning lunch time on day t, 1 otherwise;

$v2^{kt}$ Binary decision variable equals 0 if caregiver k end working later than ending lunch time on day t, 1 otherwise;

y_{ij}^{kt} Binary decision variable, equals 1 if caregiver k takes a lunch break after visiting patient i and then continue visiting patient j on day t, 0 otherwise;

Mathematical Model

$$f = Min(f_1 + f_2 + f_3 + f_4) \tag{1}$$

Subject to

$$f_1 = \sum_{i\in\Omega_P}\sum_{t\in\Omega_T} u_i^t C^P \tag{2}$$

$$f_2 = \sum_{k\in\Omega_K}\sum_{t\in\Omega_T} C^W(tE^{kt} - tB^{kt}) \tag{3}$$

$$f_3 = \sum_{k\in\Omega_K}\sum_{t\in\Omega_T} C^O \sum_{i\in\Omega_P\cup\Omega_D\cup\{0\}}\sum_{j\in\Omega_P\cup\{C\}} x_{ij}^{kt} d_{ij} \tag{4}$$

$$f_4 = C^V \sum_{k\in\Omega_K}\sum_{t\in\Omega_T}\sum_{j\in\Omega_P} x_{jC}^{kt} \tag{5}$$

$$\sum_{j\in\Omega_p} x_{0j}^{kt} + \sum_{i\in\Omega_D}\sum_{j\in\Omega_P} x_{ij}^{kt} \leq 1 \, \forall t\in\Omega_T, \forall k\in\Omega_K \tag{6}$$

$$\sum_{j\in\Omega_p} x_{ij}^{kt} \leq w_i^k \, \forall i\in\Omega_D, \forall t\in\Omega_T, \forall k\in\Omega_K \tag{7}$$

$$\sum_{j\in\Omega_p} x_{0j}^{kt} + \sum_{i\in\Omega_D}\sum_{j\in\Omega_P} x_{ij}^{kt} = \sum_{j\in\Omega_p} x_{jC}^{kt} \, \forall t\in\Omega_T, \forall k\in\Omega_K \tag{8}$$

$$\sum_{j\in\Omega_p} x_{jC}^{kt} = z^{kt}(1-z^{k(t-1)})\sum_{j\in\Omega_p} x_{0j}^{kt} + (z^{kt}(z^{k(t-1)}-1)+1)\sum_{i\in\Omega_D}\sum_{i\in\Omega_P} x_{ij}^{kt} \, \forall t\in\Omega_T \, t\neq 1, \forall k\in\Omega_K \tag{9}$$

$$\sum_{j\in\Omega_p} x_{jC}^{k1} = z^{k1}\sum_{j\in\Omega_p} x_{0j}^{k1} + (1-z^{k1})\sum_{i\in\Omega_D}\sum_{j\in\Omega_P} x_{ij}^{k1} \, \forall k\in\Omega_K \tag{10}$$

$$\sum_{i\in\Omega_P\cup\Omega_D\cup\{0\}} x_{ij}^{kt} = \sum_{i\in\Omega_P\cup\{C\}} x_{ji}^{kt} \, \forall j\in\Omega_P, \forall t\in\Omega_T, \forall k\in\Omega_K \tag{11}$$

$$u_i^t = d_i^t - \sum_{j \in \Omega_p \cup \{C\}} \sum_{k \in \Omega_K} x_{ij}^{kt} \; \forall i \in \Omega_P, \; \forall t \in \Omega_T \tag{12}$$

$$\sum_{i \in \Omega_p \cup \Omega_D \cup \{0\}} \sum_{k \in \Omega_K} x_{ij}^{kt} b P^{kq} p_j^k = n_j^{tq} \; \forall j \in \Omega_P, \; \forall t \in \Omega_T, \forall q \in \Omega_Q \tag{13}$$

$$WL^{kt} \leq t B^{kt} \; \forall t \in \Omega_T, \; \forall k \in \Omega_K \tag{14}$$

$$t E^{kt} \leq WU^{kt} \; \forall t \in \Omega_T, \; \forall k \in \Omega_K \tag{15}$$

$$t B^{kt} \leq t s_i^{kt} \leq t E^{kt} \; \forall i \in \Omega_P \cup \Omega_D \cup \{0\} \cup \{C\}, \forall t \in \Omega_T, \forall k \in \Omega_K \tag{16}$$

$$x_{ij}^{kt} P L_i^t \leq t s_i^{kt} \; \forall i \in \Omega_P, \forall j \in \Omega_P \cup \{C\}, \forall t \in \Omega_T, \forall k \in \Omega_K \tag{17}$$

$$t s_i^{kt} + (x_{ij}^{kt} - 1) \times M \leq P U_i^t \; \forall i \in \Omega_P, \forall j \in \Omega_P \cup \{C\}, \forall t \in \Omega_T, \forall k \in \Omega_K \tag{18}$$

$$t s_i^{kt} + s_i^t + t t_{ij} + (x_{ij}^{kt} - 1) \times M \leq t s_j^{kt} \; \forall i \in \Omega_P, \forall j \in \Omega_P \cup \{C\}, \forall t \in \Omega_T, \forall k \in \Omega_K \tag{19}$$

$$t s_i^{kt} + t t_{ij} + (x_{ij}^{kt} - 1) \times M \leq t s_j^{kt} \; \forall i \in \Omega_D \cup \{0\}, \forall j \in \Omega_P, \forall t \in \Omega_T, \forall k \in \Omega_K \tag{20}$$

$$t E^{kt} - t B^{kt} - \sum_{i \in \Omega_p} \sum_{j \in \Omega_p} y_{ij}^{kt} L^{break} \leq L \; \forall t \in \Omega_T, \; \forall k \in \Omega_K \tag{21}$$

$$LB \times v1^{kt} \leq t B^{kt} \leq LB + v1^{kt} \times M \; \forall t \in \Omega_T, \; \forall k \in \Omega_K \tag{22}$$

$$LE \times (1 - v2^{kt}) \leq t E^{kt} \leq LE + (1 - v2^{kt}) \times M \; \forall t \in \Omega_T, k \in \Omega_K \tag{23}$$

$$\sum_{i \in \Omega_p} \sum_{j \in \Omega_p} y_{ij}^{kt} \geq 1 - v1^{kt} - v2^{kt} \; \forall t \in \Omega_T, k \in \Omega_K \tag{24}$$

$$\sum_{i \in \Omega_p} \sum_{j \in \Omega_p} y_{ij}^{kt} \leq 1 - v1^{kt} \; \forall t \in \Omega_T, k \in \Omega_K \tag{25}$$

$$\sum_{i \in \Omega_p} \sum_{j \in \Omega_p} y_{ij}^{kt} \leq 1 - v2^{kt} \; \forall t \in \Omega_T, k \in \Omega_K \tag{26}$$

$$LB \leq t L^{kt} \leq LE \; \forall t \in \Omega_T, \; \forall k \in \Omega_K \tag{27}$$

$$t s_i^{kt} + s_i^t \leq t L^{kt} + (1 - \sum_{j \in \Omega_P} y_{ij}^{kt})(s_i^t + M) \; \forall i \in \Omega_P, \forall t \in \Omega_T, \forall k \in \Omega_K \tag{28}$$

$$tL^{kt} + L^{break} \leq ts_j^{kt} + (1 - \sum_{i \in \Omega_P} y_{ij}^{kt})(L^{break} + LE) \, \forall j \in \Omega_P, \forall t \in \Omega_T, \forall k \in \Omega_K$$

$$(29)$$

$$x_{iC}^{kt} = 0 \, \forall i \in \Omega_D \cup \{0\}, \forall t \in \Omega_T, \forall k \in \Omega_K \tag{30}$$

$$h_i - h_j + x_{ij}^{kt}(n+1) \leq n \, \forall i, j \in \Omega_P, \forall t \in \Omega_T, \forall k \in \Omega_K \tag{31}$$

$$x_{ij}^{kt} \in \{0, 1\} \, \forall i \in \Omega_P \cup \Omega_D \cup \{0\}, \forall j \in \Omega_P \cup \{C\}, \forall t \in \Omega_T, \forall k \in \Omega_K \tag{32}$$

$$u_i^t \in \{0, 1\} \, \forall i \in \Omega_P \tag{33}$$

$$y_{ij}^{kt} \in \{0, 1\} \, \forall i, j \in \Omega_P, \forall t \in \Omega_T, \forall k \in \Omega_K \tag{34}$$

$$v1^{kt}, v2^{kt} \in \{0, 1\} \, \forall k \in \Omega_K, t \in \Omega_T \tag{35}$$

Constraints (2)–(5) correspond to the four parts of the objective function respectively. Constraint (2) indicates the penalty cost for unscheduled visits where a universal penalty cost per time unit is considered for each patient that could not be served as scheduled. Constraint (3) represents wage paid to time spent by caregivers to accomplish the homecare services scheduled by the center. This part takes into account only the time spent by the caregivers to fulfill their assignments scheduled by the HHC center and no overtime work is considered. Constraint (4) shows the total oil costs which have relative with the travel distance of caregivers. Constraint (5) represents the total charges for cars. Because the daily charges for rental cars and the private cars of caregivers are assumed to be the same in this study, it is enough to calculate the total number of cars used during the planning horizon. Since each caregiver must return to the HHC center when they finish serving all patients, the total number of the routes back to the destination C equals to the number of the working vehicles.

Constraints (6)–(10) help to illustrate the topology of the network and to ensure the balance of the flows in and out of each location. Constraint (6) ensures that each caregiver departs from either home or HHC center once every planning day, or that he/she may stay at the HHC center when he/she is not assigned, so the total number of routes starting from HHC center and homes should be no more than 1. Constraint (7) ensures that the caregivers using their private cars as working vehicles depart from their own homes. Constraint (8) guarantees that each caregiver ends up the daily route at the HHC center by making the total number of both the caregivers depart from both their homes and those starting their work from the HHC center equal to the number of the caregivers that return to the HHC center after finishing their daily work. Constraints (9) and (10) guarantee that the number of caregivers returning back to the HHC center equals to the total number of caregivers departing from both the HHC center and their homes for the days t ≠ 1 and the first day (t = 1) respectively.

Constraints (11)–(13) ensure that the patient can be served on request. Constraint (11) indicates that once a caregiver visits one patient, he/she must leave this patient's location after finishing the service. Constraint (12) represents whether the patient with demand is served or not. In fact, a patient's demand of one day is not served when no route passes through the location of that patient on that day. Constraint (13) guarantees that the caregivers assigned to a patient must respect both the qualification level required and the preference of that patient.

Constraints (14)–(21) are related to time restrictions. Constraints (14) and (15) ensure that the starting and ending working time of the caregivers will be set within their contract working time. Constraint (16) represents the time restriction of starting service time point at locations. Constraints (17) and (18) ensure the feasibility of each visit according to the predefined time windows. Constraints (19) and (20) guarantee that caregivers have enough time to travel between two consecutive locations. Constraint (19) stands for the time from a patient's location to another patient's location or HHC center, while constraint (20) stands for the time from HHC center or caregiver's home to a patient's location. Constraint (21) indicates the maximum daily working duration. Since the lunch break time is paid but not considered as regular working time, the daily working time equals the gap between the ending time and the starting time of the route minus possible lunch break time. Furthermore, the working time needs to be no more than the predefined threshold.

Constraints (22)–(29) are dedicated to the lunch break. Constraints (22)–(26) indicate that a lunch break should be assigned to a caregiver if this caregiver starts the work earlier than the lunch start time and finishes working later than the lunch end time. Constraint (27) indicates that the time window of the lunch break must be respected. Constraints (28) and (29) indicate that no-preemption of the service is allowed, where constraint (28) guarantees the caregivers to keep enough time for finishing serving a patient before taking a lunch break, and constraint (29) ensures the caregivers to keep enough time for taking a lunch break before serving the next patient.

The rest of the constraints are constructed to ensure the feasibility of the final solution. Constraint (30) avoids the route establishment from any caregiver's home to the HHC center. Constraint (31) is used to eliminate sub-tours, sub-tours means that some patients connect with each other to form a tour in which starting and ending point are not included, and the constraints above obviously cannot avoid the sub-tours. More details about sub-tours elimination can be found in [35]. Constraints (32)–(35) represent the feasibility of the decision variables, respectively.

3.3 Fixed Depart Way for Caregivers

In order to help HHC center managers better make proper decision to get the optimal solution, we further discuss the relationship between the caregivers' geographical areas and optimal results. For this purpose, the other two new scenarios are assumed in which caregivers complete the visiting routes with a fixed depart way, meaning that all caregivers start from HHC center or all caregivers start from their domiciles. As a result, the model we construct in Sect. 3.1 need to be modified to meet these new constraints.

The first scenario assumes that HHC center rents the car for each caregiver for their daily work, and each caregiver needs to return the car after he/she finishes the assigned services; in this new model, the constraint (36) replaces the constraint (6), and constraints (7)–(10) and (31) are no longer needed in this model as there is no possibility for caregivers depart from their homes. The other equations are same as the Sect. 3.2.

$$\sum_{j\in\Omega_p} x_{0j}^{kt} \leq 1 \forall t \in \Omega_T, \ \forall k \in \Omega_K \tag{36}$$

The second scenario assumes that all caregivers using their own car to serve the patients, hence, all caregivers depart from their homes and back to the HHC center. Considering this new situation, in this model, the constraint (37) replaces the constraint (6), and the constraints (8) to (10) are no longer needed in this model because it is unlikely for the caregivers to depart from HHC center. The other constraints are the same as the Sect. 3.2.

$$\sum_{i\in\Omega_D} \sum_{j\in\Omega_P} x_{ij}^{kt} \leq 1 \ \forall t \in \Omega_T, \ \forall k \in \Omega_K \tag{37}$$

4   Experiment Results

Mid-term HHC planning problem has been rarely treated with various constraints before; to the best of our knowledge, there is no a similar model that exists in the literature, and hence useful benchmark instances are missing to validate the model. For this reason, we first reduce our problem into a PVRPTW problem. Classical PVRPTW instances were proposed by Cordeau in 2001 [36], and the instances range from 48 to 288 patients as well as from 3 to 20 homogeneous vehicles, and have a planning horizon of 4 or 6 days. In those 4-day instances, patients demand the visit for 4 times, 2 times and once respectively in planning horizon, while in 6-day instances, the visit in planning horizon is required by patients for 6 times, 3 times, 2 times and once respectively. In addition, considering the time window, in depot (e.g. HHC center), the time window is [0, 1000]; all time windows are within this time frame. Other parameters not included in instances are generated randomly.

Since PVRPTW has been proven to be NP-hard and the targeted problem covers all the constraints of the general PVRPTW, it is also NP-hard. In general, exact methods such as branch-and-bound, can be applied to obtain the optimal solution

Table 5 Variation domain for the simulation parameters

Parameters	Variation domain
Planning horizon	4–7
Number of caregivers	2–12
Number of patients	4–20

of an NP-hard problem, while heuristics or metaheuristics are more appropriate to solve large-size NP-hard problems for relative high quality solution within reasonable execution time. According to the literature, some commercial programming solvers, such as CPLEX and Gurobi, have been developed to enable the researchers to obtain optimal solution of the problems that can be formulated as linear programming models, and those tools are quite efficient for small-size instances. In this study, the commercial programming solver Gurobi (Gurobi 8.1.0) is applied to obtain optimal solutions for small-size instances to validate the constructed model on a laptop with Intel Core i7-8700 CPU, 32 GB RAM, 3,2 GHz and Linux system.

Four groups, containing in total 21 small-size instances, are tested in this study. Group A and B consist of classical PVRPTW instances with the planning horizon 4-day and 6-day respectively; group C and group D are completely randomly generated data with the planning horizon of 5-day and 7-day respectively. The variation domain of the simulation parameters is shown in Table 5. The penalty cost for an uncovered demand is set as 200; the payment is 0.2 per time unit (each time unit equals 1) for each caregiver; oil cost for a car is 0.1 per distance unit (each distance unit equals 1); the charge for a car is set as 12 for each day.

As mentioned above, in order to analyze the relationship between the caregivers' geographical area and objective function, three different scenarios (CDC, CC and DC) were defined for each instance by varying the departure models of the caregivers. In consequence, $21 \times 3 = 63$ scenarios are solved in total, and computational results are shown in Table 6, where column "Group" indicates the information about which groups the instances belong to, "K" represents the number of caregivers, "N" the number of patients, "T" the planning horizon, and "D" the total demands of patients during planning horizon. For example, the first instance in Group A includes 3 caregivers and 6 patients with 14 total demands during the 4-day planning horizon.

It is worth noting that when the starting points of all the caregivers' trajectories merge into the same ending node—HHC center, as the scenarios of CC mode are, the problem turns into a PVRPTW. Thus the scenarios of CC mode can be used as the benchmark, and the comparisons conducted among CC mode and the other two modes can help us illustrating the improvement of the proposed strategy when compared with the classical one.

Four criteria are used to evaluate the quality of the solutions: "Cost" represents the objective value of the constructed model, "US" indicates the number of demands uncovered during the planning period, "Gap" represents the rate of the requests not being responded to, which is defined as $US/D \times 100 \%$ "CPU" indicates the execution time measured in seconds. It is worth noting that it might be really time consuming for Gurobi to solve this problem when the size of instance grows. In consequence,

Table 6 Computational results

Group	K	N	T	D	CDC				CC				DC			
					Cost	US	Gap (%)	CPU(s)	Cost	US	Gap (%)	CPU(s)	Cost	US	Gap (%)	CPU(s)
Group A	3	6	4	14	343.35	0	0	0.86	369.53	0	0	0.57	295.18	0	0	3.18
	3	8	4	16	585.42	0	0	9.41	607.34	0	0	4.58	640.31	0	0	14.26
	6	12	4	28	1105.73	1	3.6	21.42	1183.61	1	3.6	10.72	1153.90	1	3.6	24.16
	6	13	4	31	888.079	0	0	640.32	1216.75	0	0	430.96	979.89	0	0	1259.81
	9	15	4	33	1071.60	0	0	150.66	1228.07	0	0	860.16	1200.14	0	0	292.85
	9	16	4	34	965.724	0	0	595.76	1073.90	0	0	1153.14	1232.05	0	0	534.22
	12	18	4	37	1056.70	0	0	470.88	1230.72	0	0	1913.60	1189.90	0	0	2970.23
	12	20	4	41	1228.76	0	0	1723.08	1357.49	0	0	2876.47	1346.72	0	0	1305.93
Group B	4	6	6	17	635.116	0	0	23.36	689.55	0	0	17.01	414.43	0	0	16.51
	4	7	6	19	475.38	0	0	29.11	557.84	0	0	35.13	519.53	0	0	107.59
	5	10	6	30	796.22	0	0	326.83	940.77	0	0	373.45	903.21	0	0	1017.61
	5	12	6	35	1289.75	1	2.9	29.07	1404.05	1	2.9	14.074	1293.24	1	2.9	31.78
	8	14	6	43	1405.98	0	0	440.92	1621.06	0	0	1099.68	1475.79	0	0	509.85
	8	15	6	47	1416.41	0	0	150.56	1624.61	0	0	761.77	1661.60	0	0	193.13
Group C	4	8	5	18	428.99	0	0	20.11	465.95	0	0	29.97	441.92	0	0	9.26
	4	10	5	23	528.77	0	0	507.73	554.06	0	0	2245.80	547.72	0	0	577.96
	7	13	5	35	–	–	–	$>1\times10^4$	–	–	–	$>1\times10^4$	–	–	–	$>1\times10^4$
Group D	2	4	7	18	592.87	0	0	1.64	653.34	0	0	0.26	641.12	0	0	0.28
	2	5	7	21	559.10	0	0	3.03	630.37	0	0	0.51	608.06	0	0	0.52
	6	10	7	42	1729.58	0	0	9.28	1987.53	0	0	10.23	1889.58	0	0	4.90
	6	12	7	44	1595.53	0	0	8.74	1760.76	0	0	17.51	1744.09	0	0	10.01

Table 7 Evaluated results of three models

Model	Proportion of optimal cost	Proportion of least computational time
CDC	18/20 = 90%	8/20 = 40%
CC	0/20 = 0%	7/20 = 35%
DC	2/20 = 10%	5/20 = 25%

the maximum execution time is set as 10000 s in this study to make Gurobi stop working and report the final result when some instances are too time consuming.

As shown in Table 6, 20 instances (95%) can be solved by Gurobi within the limit time; only the last instance in Group C cannot report the optimal solution. For the 20 instances with optimal solution, a huge gap is observed with regards to the computational time. For example, in the model CDC, the shortest CPU time is only 0.86 s (the first instance in Group A), while it takes 1723.08 s to solve the last instance in the same group. Nevertheless, even the longest execution time is acceptable according to the professionals working for the HHC center.

As is shown in Table 6, the results of "US" and "Gap" are the same between three models of each instance, so we mainly compared the results with objective value and the computational time. Also we take the first instance in Group A as an example: the optimal cost of model CDC is 343.35 with 0.86 s consumed time needed while the model CC obtained 369.532 as the objective function value and the computing time is 0.57 s; and it takes 3.18 s for model DC to get optimal cost 295.18. Therefore, in this instance, model DC is the best model because of the lowest cost although it takes longer time to obtain the optimal solution.

In order to deeply evaluate the three proposed model, a comparative analysis about the number of instances with optimal cost and least computational time for each model is presented in Table 7; the results show that CDC is the most optimal model for obtaining optimal solution for 18 instances (90%) whereas CC cannot even obtains the optimal solution, and only 2 instances (10%) of Model DC are solved with optimality. As for the time consumption, the model CDC also shows the highest efficiency with 8 instances (40%) that reach the optimal with least computational time, and the number of instances with least computational time for CC and DC are 7 (35%) and 6 (25%), respectively.

Furthermore, since the cost consists of four parts (mentioned in Sect. 3.2), the detail cost of each instance is presented in Table 8, where column "Cost1" represents the penalty for unscheduled visits, "Cost2" the wages of caregivers, "Cost3" the fuel cost and "Cost4" the charges for operational vehicles.

The variation of costs, sub-costs, and computational time of 20 instances under three models can be seen in Figs. 3, 4, 5. Figure 3 shows that three models have the same trends between different instances and costs, all the minimum costs take place in instance k3n6t4, and k6n10t7 is the instance that keeps maximum cost in three models. But the exact costs are different when it comes to each model, we can see that model CC has the highest average cost, in this model, minimum cost is 343.35, and maximum cost is 1729.58. The lowest average cost belongs to model CDC, in

46   W. Liu et al.

Table 8 Computational results of sub-costs

Group	K	N	T	D	CDC				CC				DC			
					Cost1	Cost2	Cost3	Cost4	Cost1	Cost2	Cost3	Cost4	Cost1	Cost2	Cost3	Cost4
Group A	3	6	4	14	0	198.51	84.84	60	0	212.76	96.77	60	0	161.57	73.61	60
	3	8	4	16	0	322.96	154.46	108	0	373.36	137.98	96	0	387.33	144.98	108
	6	12	4	28	0	545.69	264.04	276	0	617.21	290.40	276	0	477.40	400.50	276
	6	13	4	31	0	459.81	224.27	204	0	708.66	304.09	204	0	544.62	231.27	204
	9	15	4	33	0	586.99	256.61	228	0	700.80	287.27	240	0	684.25	287.89	228
	9	16	4	34	0	470.69	231.03	264	0	621.00	224.90	228	0	666.85	277.20	288
	12	18	4	37	0	526.05	254.65	276	0	711.25	291.47	228	0	634.56	267.34	288
	12	20	4	41	0	628.84	311.92	288	0	728.34	341.15	288	0	711.28	347.44	288
Group B	4	6	6	17	0	363.30	163.82	108	0	423.43	194.12	72	0	237.31	105.12	72
	4	7	6	19	0	251.22	116.16	108	0	321.62	140.22	96	0	295.76	127.77	96
	5	10	6	30	0	424.69	164.53	180	0	544.58	216.19	180	0	483.59	211.62	180
	5	12	6	35	200	544.90	256.85	288	200	635.43	280.62	288	200	564.32	240.92	288
	8	14	6	43	0	713.14	332.84	360	0	885.58	399.48	336	0	773.26	342.53	360
	8	15	6	47	0	705.64	338.76	372	0	882.14	394.47	348	0	873.86	391.74	396
Group C	4	8	5	18	0	243.67	113.32	72	0	278.43	127.52	60	0	270.41	111.51	60
	4	10	5	23	0	307.07	137.7	84	0	326.57	143.49	84	0	406.35	141.37	84
	7	13	5	35	–	–	–	–	–	–	–	–	–	–	–	–
Group D	2	4	7	18	0	348.95	159.92	84	0	398.16	171.18	84	0	390.01	167.11	84
	2	5	7	21	0	348.78	126.32	84	0	380.58	165.79	84	0	365.70	158.36	84
	6	10	7	42	0	880.90	404.68	444	0	1068.95	474.58	444	0	1003.65	429.93	456
	6	12	7	44	0	753.17	374.36	468	0	902.71	390.05	468	0	883.59	380.50	480

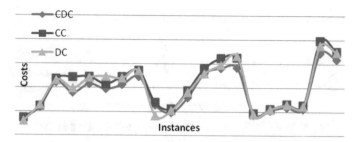

Fig. 3 The variation of costs depending on instances

this model, the minimum cost is 369.53, and the maximum cost is 1987.53. For model DC, the minimum and maximum costs are 295.18 and 1889.58 respectively. Figure 4 presents the variation of four sub-costs depending on different instances, an obvious situation can be observed in Fig. 4a that the costs of unscheduled visits are the same among three models; this can also refer in the results of "US" and "Gap", the other three figures of Fig. 4 show that three models keeps almost same trends of variation between Cost2, 3 and 4 and different instances. In Fig. 5, the trends between computational time and instances also show consistency among three models, but this tendency is not as obvious as the trend of costs.

Moreover, as Table 6 indicates that optimal results are influenced by the characters: the number of caregivers, number of patients, number of patients' demands and duration of planning horizon. Generally, the number of patients' demands is the most important factor that affects the objective function value; the more demands there are, the more services need to be assigned, and the more expenditures is incurred.

The relationship between the number of patient's demands and the objective cost is presented in Fig. 6. It is worth noting that since certain instances keep the same demands, the average of the cost is calculated. For example, there are two instances with 18 demands, and the cost for 18 demands is obtained by: $(428.99 + 592.87)/2 = 510.93$ for the model CDC, $(465.95 + 653.34)/2 = 559.65$ for the model CC, and $(441.92 + 641.12) = 541.52$ for the model DC. As shown in Fig. 6, an overall trend can be observed: while the demands increase, the objective value increases as well for all three models, but there is no a clear function to model the variation of objective function value and number of demands.

Figure 7 shows how four instances characters influence computational time. Similar to the analysis of cost, the average of computational times is calculated for each number of characters. Figure 7a indicates that the increase in the computational time with the increase in the number of caregivers, and it is also noted that as the number of patients exceeds 9, computational time will increase dramatically. Figure 7b presents that there is a positive correlation between the calculation time and the number of caregivers, and also we can find that model CC keeps the weakest correlation in these three models. Figure 7c shows a negative correlation between the calculation time and the planning horizon duration, and this might be due to the fact that the larger the planning horizon becomes, the lower the number of daily demands come out.

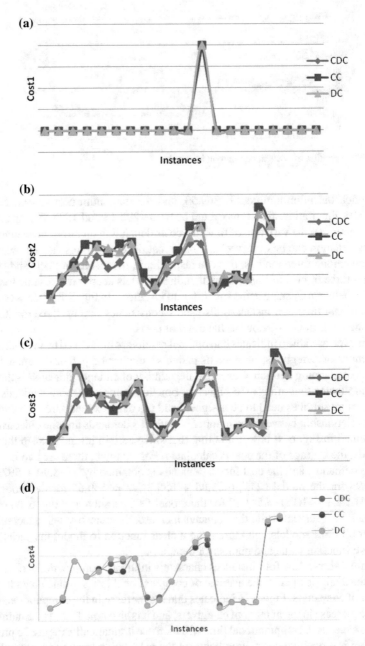

Fig. 4 **a** The variation of cost1 depending on instances **b** The variation of cost2 depending on instances **c** The variation of cost3 depending on instances **d** The variation of cost4 depending on instances

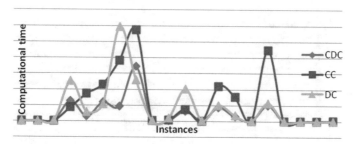

Fig. 5 The variation of computational time depending on instances

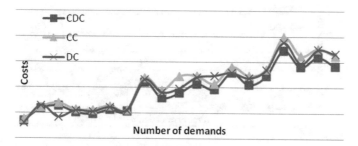

Fig. 6 Costs under different number of demands

Finally in Fig. 7d, it is obvious that the variation of the computational time does not follow a specific function of the number of demands.

Furthermore, we show the detail of optimal results of instances with 6 caregivers, 12 patients and 4-day planning horizon solved by each model in Table 9. The column "Model" indicates the type of model we used; the next two columns are the IDs of the planning days and the caregivers who will be dispatched on these planning days. Routes that caregivers need to complete to serve patients are named as "Route"; "WL" and "WU" are the earliest and latest working time of each caregiver, while "TB" and "TD" mean the caregiver's beginning and ending time point of a route, respectively. The caregiver's time utilization rate is showed in column "TW", while equals $(WU - WL)/(TD - TB) \times 100\%$, and indicates the workload of different caregivers on their working days. The next column "Avg.TW" shows the average time utilization rate of all caregivers under three models, while model CC shows the best performance with the average time utilization rate maximized, but correspondingly, model CDC keeps the minimum time utilization rate. The column "RR" demonstrates the response rate by HHC center that corresponds to the patient's daily demands; it is noted that three model show the same results that on the fourth day, a patient's demand cannot be satisfied on the fourth day with $RR = (7/8) \times 100\% = 87.5\%$. The last column "Break" determines whether a lunch break happens, and value 1 means caregivers need to take a lunch break; 0 otherwise. Figure 8 illustrates the optimal routes of this instance on the third day under three models, on this day, Fig. 8a shows that if we take model CDC, there are four caregivers dispatched among whom the caregivers

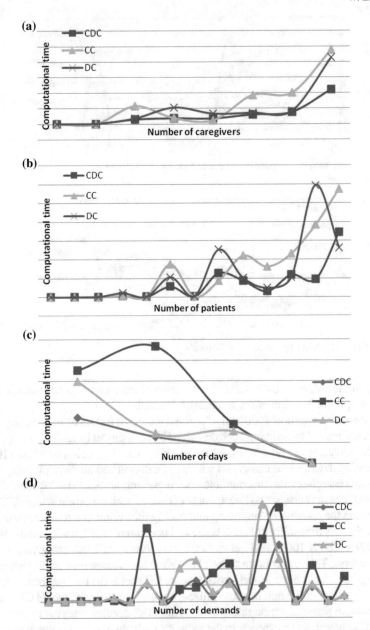

Fig. 7 **a** Computational time under different number of caregivers **b** Computational time under different number of patients **c** Computational time under different number of days **d** Computational time under different number of demands

Table 9 Optimal results of instance k6n12t4

Model	Days	Caregivers	Route	WL	WU	TB	TD	TW (%)	Avg.TW (%)	RR (%)	Break
CDC	1	1	13-4-C	200	700	374	462	17.6	28.6	100	0
		2	14-3-2-1-C	300	750	357	538	40.2			1
		4	0-9-5-C	100	850	315	441	16.8			0
		6	0-8-C	50	650	161	198	6.16			0
	2	2	0-3-2-1-C	300	750	412	569	34.8		100	1
		3	15-10-6-4-C	200	800	230	485	42.5			0
	3	2	14-3-1-C	300	750	272	408	30.2		100	0
		3	0-8-4-C	200	800	292	441	24.8			0
		5	17-7-C	200	600	219	385	41.5			0
		6	18-11-2-C	50	650	321	490	28.1			0
	4	1	13-12-C	200	700	273	363	18		87.5	0
		2	14-3-2-1-C	200	650	357	538	40.2			1
		3	15-6-4-5-C	200	800	281	468	31.2			0
CC	1	1	0-5-2-C	200	700	374	492	23.6	33.2	100	0
		3	0-8-3-9-C	200	800	236	491	42.5			0
		6	0-4-1-C	50	650	308	460	24.9			0
	2	3	0-10-4-C	200	800	200	481	46.8		100	0
		4	0-6-3-C	100	850	290	468	23.7			0
		5	0-1-2-C	200	600	382	542	37.5			1
	3	2	0-1-7-C	300	750	342	560	48.4		100	1
		4	0-3-11-2-C	100	850	303	472	28.1			0

(continued)

Table 9 (continued)

Model	Days	Caregivers	Route	WL	WU	TB	TD	TW (%)	Avg.TW (%)	RR (%)	Break
		6	0-4-8-C	50	650	200	332	22			0
	4	2	0-4-12-C	300	750	345	562	48.2		87.5	1
		3	0-1-6-3-C	200	800	300	478	29.7			0
		6	0-5-2-C	50	650	509	646	22.8			0
DC	1	1	13-5-C	200	700	417	482	13	29.3	100	0
		4	16-3-9-C	100	850	355	567	28.2			1
		5	17-2-1-8-4-C	200	600	276	540	66			1
	2	1	13-1-0	200	700	355	417	12.4		100	0
		2	14-2-0	300	750	418	489	15.7			0
		3	15-10-4-C	200	800	200	441	40.1			0
		4	16-6-3-C	100	850	320	478	21.1			0
	3	2	14-1-8-C	300	750	355	459	22.4		100	0
		3	15-3-4-C	200	800	236	341	17.5			0
		5	17-7-2-11-C	200	600	219	385	41.5			0
	4	3	15-2-1-5-C	200	800	541	789	41.3		87.5	0
		4	16-6-3-C	100	850	242	400	21.1			0
		5	17-4-12-0	200	600	277	440	40.8			0

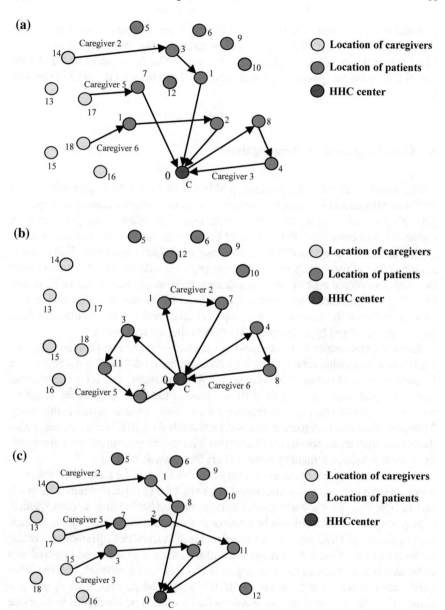

Fig. 8 a Optimal routes of instance k6n12t4 on the third day under model CDC **b** Optimal routes of instance k6n12t4 on the third day under model CC **8 c** Optimal routes of instance k6n12t4 on the third day under model DC

2, 5 and 6 depart from their homes and the caregiver 3 departs from HHC center. Figure 8b indicates that three caregivers 2, 4 and 6 are assigned with departure from HHC center if model CC is taken. The last image Fig. 8c presents that under the model DC, three caregivers 2, 3, 5 will serve the demands with departure from their homes.

5 Conclusions and Perspectives

A mid-term home health care planning problem is studied and a novel mathematical model is constructed by taking into account a set of real-life constraints. From the patient's point of view, we take into account their time window and preferences of caregivers. For caregivers, the contracted time, qualification level and lunch break, as well as the less concerned flexible departure modes are considered. This problem is modeled as a linear integer programming problem with an objective to minimize the total operational cost of home health care services offered by the HHC center. Furthermore, the other two scenarios CC and DC are proposed to discuss the relationship between the caregivers' geographical areas and optimal results. All these models are validated by a commercial programming solver Gurobi.

Experimental results show that most of the small-size instances can be solved by Gurobi in a reasonable time with high quality, and the model CDC is the best model because it can find the optimal solution for 90% instances, and solve the problem with least computational time for 40% instances. Model CC gets the worst results with no optimal solution can be obtained for all those instances tested in this study. Therefore, the flexible departure way is most suitable for a HHC center manager with the lowest cost. It can also be concluded that with the increase in patients' demands, the objective value obtained by three models increases as well.

It is worth noting that Gurobi is not very efficient to when the scale of the instances becomes large because it was mentioned before that the targeted problem is NP-hard, and that Gurobi cannot solve large-size instances within limited time. In consequence, high quality meta-heuristics will be developed in the near future. In addition, in practices operation of HHC centers, as the optimal results show that differences between caregiver's time utilization rates exist, workload balancing need to be involved as a constraint in the model in the next step to keep caregivers motivated. One the other hand, many uncertain aspects exist in HHC operational process, such as unavailability of caregivers and patients, traveling time between two locations and service duration [37, 38]. In our mid-term home health care planning problem, patient's demand might be canceled temporarily or some urgent demands need to be added during the horizon, and this problem will be changed into a stochastic problem; considering such constraint, a more complicated stochastic model needs to be devised in the future.

Acknowledgements The first author thanks the China Scholarship Council for financial support gratefully. (Contract N.201801810101).

References

1. Sahoo, A.K., S. Mallik, C. Pradhan, B.S.P. Mishra, R.K. Barik, and H. Das. 2019. Intelligence-based health recommendation system using big data analytics. In *Big data analytics for intelligent healthcare management,* 227–246. Academic Press.
2. Emiliano, W., J. Telhada, and M. Carvalho. 2017. Home health care logistics planning: A review and framework. *Procedia Manufacturing.* 13: 948–955.
3. Harris-Kojetin L., M. Sengupta, E. Park-Lee, and R. Valverde. 2013. Long-term care services in the United States: 2013 overview. National Center for Health Statistics Vital Health Stat, 3–37.
4. Morris, M. 2016. Global health care outlook: Battling costs while improving care.
5. Genet, N., W. Boerma, M. Kroneman, A. Hutchinson, and R.B. Saltman. 2012. *Home care across Europe—current structure and future challenges. European observatory on health systems and policies.* Oslo, Norway: World Health Organization.
6. Mankowska, D., F. Meisel, and C. Bierwirth. 2013. The home health care routing and scheduling problem with interdependent services. *Health Care Management Science* 17: 15–30.
7. Begur, S., D. Miller, and J. Weaver. 1997. An integrated spatial DSS for scheduling and routing home-health-care nurses. *Interfaces* 27: 35–48.
8. Cheng, E., and J.L. Rich. 1998. A home health care routing and scheduling problem. Department of CAAM, Rice University, Houston Texas.
9. Bektas, T. 2006. The multiple traveling salesman problem: An overview of formulations and solution procedures. *Omega* 34: 209–219.
10. Akjiratikarl, C., P. Yenradee, and P. Drake. 2007. PSO-based algorithm for home care worker scheduling in the UK. *Computers & Industrial Engineering* 53: 559–583.
11. Kergosien Y., C. Lenté, and J-C. Billaut. 2009. Home health care problem: An extended multiple traveling salesman problem. In *4th multidisciplinary international conference on scheduling: Theory and applications,* Dublin, Ireland.
12. Trautsamwieser, A., M. Gronalt, and P. Hirsch. 2011. Securing home health care in times of natural disasters. *OR Spectrum* 33: 787–813.
13. Eveborn, P., P. Flisberg, and M. Rönnqvist. 2006. Laps Care—an operational system for staff planning of home care. *European Journal of Operational Research* 171: 962–976.
14. Nickel, S., M. Schröder, and J. Steeg. 2012. Mid-term and short-term planning support for home health care services. *European Journal of Operational Research* 219: 574–587.
15. Rodriguez, C., T. Garaix, X. Xie, and V. Augusto. 2015. Staff dimensioning in homecare services with uncertain demands. *International Journal of Production Research* 53: 7396–7410.
16. Decerle, J., O. Grunder, A. Hajjam El Hassani, and O. Barakat. 2019. A memetic algorithm for multi-objective optimization of the home health care problem. *Swarm and Evolutionary Computation* 44: 712–727.
17. Wirnitzer, J., I. Heckmann, A. Meyer, and S. Nickel. 2016. Patient-based nurse rostering in home care. *Operations Research for Health Care* 8: 91–102.
18. Braekers, K., R. Hartl, S. Parragh, and F. Tricoire. 2016. A bi-objective home care scheduling problem: Analyzing the trade-off between costs and client inconvenience. *European Journal of Operational Research* 248: 428–443.
19. Xiao, L., M. Dridi, and A. El Hassani. 2018. Mathematical model for the home health care scheduling and routing problem with flexible lunch break requirements. *IFAC-PapersOnLine* 51: 334–339.
20. Bachouch, R., A. Guinet, and S. Hajri-Gabouj. 2011. A decision-making tool for home health care nurses' planning. *Supply Chain Forum: An International Journal* 12: 14–20.
21. Liu, R., B. Yuan, and Z. Jiang. 2016. Mathematical model and exact algorithm for the home care worker scheduling and routing problem with lunch break requirements. *International Journal of Production Research* 55: 558–575.
22. Chen, X., B. Thomas, and M. Hewitt. 2017. Multi-period technician scheduling with experience-based service times and stochastic customers. *Computers & Operations Research* 82: 1–14.

23. Moussavi, S., M. Mahdjoub, and O. Grunder. 2019. A matheuristic approach to the integration of worker assignment and vehicle routing problems: Application to home healthcare scheduling. *Expert Systems with Applications* 125: 317–332.
24. Bennett, A., and A. Erera. 2011. Dynamic periodic fixed appointment scheduling for home health. *IIE Transactions on Healthcare Systems Engineering* 1: 6–19.
25. Maya Duque, P., M. Castro, K. Sörensen, and P. Goos. 2015. Home care service planning. The case of Landelijke Thuiszorg. *European Journal of Operational Research* 243: 292–301.
26. Liu, R., X. Xie, and T. Garaix. 2014. Hybridization of tabu search with feasible and infeasible local searches for periodic home health care logistics. *Omega* 47: 17–32.
27. Bard, J., Y. Shao, and A. Jarrah. 2013. A sequential GRASP for the therapist routing and scheduling problem. *Journal of Scheduling* 17: 109–133.
28. Hewitt, M., M. Nowak, and N. Nataraj. 2016. Planning Strategies for Home Health Care Delivery. *Asia-Pacific Journal of Operational Research* 33: 1650041.
29. Hiermann, G., M. Prandtstetter, A. Rendl, J. Puchinger, and G. Raidl. 2013. Metaheuristics for solving a multimodal home-healthcare scheduling problem. *Central European Journal of Operations Research* 23: 89–113.
30. Redjem, R., and E. Marcon. 2015. Operations management in the home care services: A heuristic for the caregivers' routing problem. *Flexible Services and Manufacturing Journal* 28: 280–303.
31. Trautsamwieser, A., and P. Hirsch. 2014. A branch-price-and-cut approach for solving the medium-term home health care planning problem. *Networks* 64: 143–159.
32. Ikegami, A., and A. Uno. 2007. Bounds for staff size in home help staff scheduling. In *The 50th Anniversary of the Operations Research Society of Japan* 4: 563–575.
33. Fikar, C., and P. Hirsch. 2017. Home health care routing and scheduling: A review. *Computers & Operations Research* 77: 86–95.
34. Cissé, M., S. Yalçındağ, Y. Kergosien, E. Şahin, C. Lenté, and A. Matta. 2017. OR problems related to Home Health Care: A review of relevant routing and scheduling problems. *Operations Research for Health Care* 13: 1–22.
35. Kara, I., G. Laporte, and T. Bektas. 2004. A note on the lifted Miller–Tucker–Zemlin subtour elimination constraints for the capacitated vehicle routing problem. *European Journal of Operational Research* 158: 793–795.
36. Cordeau, J., G. Laporte, and A. Mercier. 2001. A unified tabu search heuristic for vehicle routing problems with time windows. *Journal of the Operational Research Society* 52: 928–936.
37. Shi, Y., T. Boudouh, O. Grunder, and D. Wang. 2018. Modeling and solving simultaneous delivery and pick-up problem with stochastic travel and service times in home health care. *Expert Systems with Applications* 102: 218–233.
38. Shi, Y., T. Boudouh, and O. Grunder. 2019. A robust optimization for a home health care routing and scheduling problem with consideration of uncertain travel and service times. *Transportation Research Part E: Logistics and Transportation Review* 128: 52–95.

Performance Analysis of NASNet on Unconstrained Ear Recognition

K. Radhika, K. Devika, T. Aswathi, P. Sreevidya, V. Sowmya and K. P. Soman

Abstract *Recent times are witnessing greater influence of Artificial Intelligence (AI) on identification of subjects based on biometrics. Traditional biometric recognition algorithms, which were constrained by their data acquisition methods, are now giving way to data collected in the unconstrained manner. Practically, the data can be exposed to factors like varying environmental conditions, image quality, pose, image clutter and background changes. Our research is focused on the biometric recognition, through identification of the subject from the ear. The images for the same are collected in an unconstrained manner. The advancements in deep neural network can be sighted as the main reason for such a quantum leap. The primary challenge of the present work is the selection of appropriate deep learning architecture for unconstrained ear recognition. Therefore the performance analysis of various pretrained networks such as VGGNet, Inception Net, ResNet, Mobile Net and NASNet is attempted here. The third challenge we addressed is to optimize the computational resources by reducing the number of learnable parameters while reducing the number of operations.* Optimization of selected cells as in NASNet architecture is a paradigm shift in this regard.

Keywords Deep learning · Convolutional neural network · Unconstrained ear recognition · NASNet

1 Introduction

Artificial Intelligence (AI) wraps the cycle, which contains Machine Learning (ML) and Deep Learning (DL) algorithms. The trend towards DL has increased day by day due to the large amount of data and high computational speed. The main advantage of DL over ML is elimination of the hard core feature extraction techniques. DL,

K. Radhika (✉) · K. Devika · T. Aswathi · P. Sreevidya · V. Sowmya · K. P. Soman
Centre for Computational Engineering and Networking Amrita School of Engineering,
Coimbatore Amrita Vishwa Vidyapeetham, Coimbatore, India
e-mail: k_radhika@cb.students.amrita.edu

© Springer Nature Switzerland AG 2020
M. Rout et al. (eds.), *Nature Inspired Computing for Data Science*,
Studies in Computational Intelligence 871,
https://doi.org/10.1007/978-3-030-33820-6_3

the subset of ML can learn from both unstructured and unlabeled data. According to Bengio [1], DL algorithms seek to exploit the unknown structure in the input distribution. This is to identify good representation often at multiple levels with higher-level learned features defined in terms of lower-level features. DL has brought about a explosion of data in all forms and from every region of the world by redefining the technological environments [2]. It has miscellaneous applications in the fields of signal processing, speech processing and synthesis, image processing, communication, optimization etc., as it discovers intricate structure in large data sets. For the past few decades, DL has shown a prominent role in computer vision problems such as image classification, object detection, object segmentation, image reconstruction etc. This chapter mainly focuses on image classification problems specific to biometric applications. These biometric applications are widely used for biomedical areas and in security purposes. One of the major issues faced in the modern world is the insecurity of personal information or any other types of data. As the usage of passwords and tokens becomes highly insecure or forgotten, biometric recognition became popular [3]. Each person has their own physical and behavioural characteristics therefore, biometric applications make use of these characteristics for person identification. Some of the biometrics such as ear, face, speech and signature can be collected unknowingly while others like iris, fingerprint, palm-print are collected with person's knowledge. Among all, finger print authentication is one of the famous and oldest biometric technique but, new studies claims that ear authentication is able to give comparatively good results to the situations in which, other authentication fails [4]. The factors like low cost, complexity, long distance recognition, static size etc., gives additional advantage to ear biometrics. Moreover, in distinguishing identical twins the difficulties with face recognition biometrics can be solved using ear biometrics [5]. However, recognition of biometrics in unconstrained environment is more challenging when compared to constrained environment.

Inorder to recognize ears in unconstrained environment, a challenge was conducted for the first time in 2017 by IEEE/IAPR, International Joint Conference on Biometrics (IJCB) [6]. Second series of group benchmarking effect (UERC-2019) was done for ear recognition by IAPR, International Conference on Biometrics 2019. There were two dataset for UERC 2019 challenge: one publically available dataset and one sequestered dataset (available only for the participants in the challenge) in which, the public dataset contains 11,000 images of 3,690 classes. Major contribution of the public dataset was carried out by Extended Annotated Web Ears (AWEx) and the remaining data were taken from UERC-2017 challenge. Sequestered dataset contains 500 images, belongs to 50 subjects which was used only for testing purpose in the challenge [7]. The trend towards DL can also be seen in biometric applications. It is observed that various Convolutional Neural Network (CNN) architectures could perform better as compared to other hand crafted feature models in UERC challenges [8].

1.1 Study on Foregoing CNN Architectures

From the year 1998, neural networks has shown an exponential growth due to the impact of two major developments in advancement of technology. During these days, data generated from various resources like social networks, cameras, mobile phones etc., has been escalated drastically. This growth demanded immense computational power inorder to process the data within a short span of time. This motivated trend towards data analysis task in various applications. DL algorithms necessitate massive amount of data to work efficiently irrespective of domain. Since DL algorithms can learn by itself from the data provided, the feature extraction task became effortless. DL has it's own supervised learning approaches such as Artificial Neural Network (ANN), CNN, Recurrent Neural Network (RNN) etc. For pattern and image recognition problems CNN has mould a prominent role [9]. Inorder to extract features automatically, CNN filters are convolved over original data. Convolutional layers, pooling layers and fully connected layers are the basic building blocks of CNN architecture. With the increase or decrease of these hidden layers, different CNN models were evolved. Evolution of most commonly used CNN architectures are shown in Table 1 [8].

In 1990s LeNet, a seven layer CNN architecture was introduced, but due to the lack of computation and memory facilities, the architecture was away from practical use till 2010. The concept of back propagation was primarily put forth by LeNet [10]. AlexNet, a much wider and deeper network than LeNet was proposed for visual object recognition tasks. Upon all classical machine learning algorithms, AlexNet attained better accuracy for recognition and is also capable of solving challenging ImageNet problems [11]. An elongation of AlexNet–ZFNet, with kernel size of 7×7, aids

Table 1 Evolution of CNN architectures

Sl.no	CNN architecture	Year	Developed by
1	LeNet	1998	Yann LeCun
2	AlexNet	2012	Alex Krizhevsky
3	ZFNet	2013	Matthew Zeiler and Rob Fergus
4	Network in network	2013	MinLin
5	VGGNet	2014	Simonyan and Zisserman
6	GoogLeNet or InceptionNet	2014	Google
7	ResNet	2015	Kaiming He
8	FractalNet	2016	G Larsson
9	DenseNet	2017	Gao Huang
10	MobileNet	2017	Google
11	XceptioNet	2017	Francois Chollet
12	NASNet	2017	Google
13	CapsuleNet	2017	Hinton

to break the raised count of weights, there upon reduced number of parameters and improved accuracies [12]. Network in Network came up with a new concept of multi layer perception convolution technique with 1×1 filters for the first time. Along with this, the network offers a technique called Global Average Pooling (GAP), which conveniently reduces the number of learnable parameters [13].

As researcher's passion towards visual recognition tasks using CNN has emerged, a new challenge called ImageNet Large Scale Visual Recognition Challenge (ILSVRC) was launched. There after, the outperformers of this challenge give rise to evolution of new models. A filter of size 3×3 was used in all the versions of VGGNet. Different versions of VGG (VGG11, VGG16 and VGG19) was developed with layers 11, 16 and 19 respectively. Each version of VGG has 8, 13 and 16 number of convolutional layers [14]. The salient feature (Inception layers) of InceptionNet or GoogLeNet made it winner of ILSVRC 2014. The advantage of GoogLeNet over other networks was the reduced number of parameters (7 M) with minor computations (1.53 G) [15]. All the evolved models concentrated on increasing the depth of the network inorder to improve accuracy. This came up with a problem of vanishing gradient. While back propagating, the value of gradient becomes too small, which does not make any sense and results in poor learning. ResNet came up with a solution for the above mentioned problem, which could increase the performance of the model. There was an exponential growth in the number of layers in various versions of ResNet [16]. An advanced version of ResNet named FractalNet arised with a new concept of drop path, which is capable of developing large models [17]. By connecting the predecessor layer output to the successive layer, a dense network (DenseNet) has formed for feature reuse [18]. For reducing the number of parameters, a new approach called separable-convolution, which includes depth-wise and point-wise separable-convolution was proposed in MobileNet. Thus, DL conquered the world of mobile and embedded applications too [19]. Standard inception modules were replaced with depth-wise separable-convolutions, which bring out another model called XceptionNet with minimum weight serialization [20]. Two promising concepts named AutoML and Neural Architectural Search (NAS) induce a new optimized architecture called NASNet. The concept of Reinforcement Learning (RL) and Evolutionary Algorithms (EAs) facilitate for the optimization task [21]. A recent model- CapsuleNet [22] on face detection has evolved for face recognition by relatively considering all the distinct facial features [23].

On the basis of the knowledge obtained from the above mentioned architectures, an elaborated study was performed for five major architectures for unconstrained ear recognition. The performance was compared by giving attention to the special features highlighted for each architecture with a special focus on the NASNEt architecture. NASNet is an optimized selection of layers which give us most satisfying results on unconstrained ear recognition. The chapter is organized as follows: Sect. 2 discusses about five DL architectures that we are discussing here. It includes VGG-16, Inception Net, ResNet, Mobile Net and NASNet. The architectural specialities are highlighted here. A detailed explanation about UERC-2019 dataset is presented in Sect. 3. Section 4 presents the design of various experiments carried out and Sect. 5 presents analysis of the results with necessary image and plots. The chapter concludes with the inference from the comparative results and its future directions.

2 CNN Architectures

Architectural details of VGG-16, MobileNet, InceptionNet, ResNet and NASNet are discussed in this section. The deep neural architectures considered here except the NASNet are developed by human experts. NASNet on the other hand is formed automatically by optimizing the individual cells in it.

2.1 Parameter Calculation for CNN Models

Convolution, max-pooling and separable-convolution are the common operations found in CNN architectures. There is a unique way of calculating number of parameters and output size for each operation. Input size, filter size, zero padding and stride are the parameters used for calculating output size of the next block. General equation for the calculation of output size is given in Eq. (1).

$$Outputsize = \frac{Inputsize - Filtersize + 2 \times zeropadding}{Stride} + 1 \quad (1)$$

As shown in Fig. 1, an RGB ($M = 3$) image has taken of size 55×55. It is convolved with 5 filters of size 5×5 by stride of 2. So, using equation [1], the output size of next block become 26×26. Max-pooling reduces the input block size by 2, thus we get 13×13 image output. While moving to separable-convolution with zero padding and stride 2 the output size become 7×7.

The general formula for calculating number of learnable parameters by convolution and separable-convolution is given in Eqs. (2) and (3) respectively.

$$No. of Learnable parameters (convolution) = N \times M \times D_k^2 \quad (2)$$

$$No. of Learnable parameters (separable - convolution) = M(D_k^2 + N) \quad (3)$$

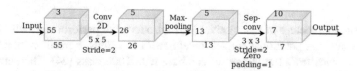

Fig. 1 Sample representation of CNN architecture

Table 2 Sample calculation of learnable parametres and output size

Operations	No. of learnable parameters	Output size
Input	0	55 × 55
Conv2D	3 × 5 × 5 × 5 = 375	26 × 26
Max-pooling	0	13 × 13
Sep-conv	(5 × 3 × 3) + (5 × 10) = 95	7 × 7

For normal convolution method, parameters are calculated by multiplying number of input channels (M) with number of filters (N) and filter size ($D_k{}^2$). From Table 2, we could observe separable-convolution reduces the number of learnable parameters. Thus we could see the phenomenon of using separable-convolution in the evolution of CNN models.

2.2 VGG16

Operations in VGG are much similar to that of AlexNet. Since VGGNet has a publically available weight configuration, it is used as baseline for image feature extraction problems. The VGGNet shown in Fig. 2, consists of stacked layers of convolution and pooling layers [14]. It can be observed that the depth of the network is 16 layers without the problem of vanishing gradients or exploding gradients. The number of learnable parameters in VGG16 is found to be 40,96,000 of 1000 class problems on ImageNet applications with 224 × 224 sized RGB images. There are 13 convolutional layers with 5 max-pooling layers and final 3 dense layers with two layers of size 4096. The size of the convolution output matrix reduces to 7 × 7 just before connecting as a dense layer. There is a fixed stride of 1, used throughout the network in convolution layers and a stride of 2, used in max-pooling layers. All the hidden layers of VGG16 uses non-linear ReLU activation functions and final layer is a softmax layer. The problem with this simple stacked layer architecture is that, it was extremely slow to train and consumes huge memory space. Varying the depth from 16–19 layers, were able to come up with better accuracy.

2.3 InceptionV3

Inception architecture of GoogLeNet [15] introduced to perform well with limited memory and computational budget. InceptionV3 is a CNN, trained on more than a million images from the ImageNet database with 1000 class [24]. The network is 48 layers deep and it has 23,885,392 learnable parameters on 1000 class problem. In order to maximize the information flow into the network, Inception module has

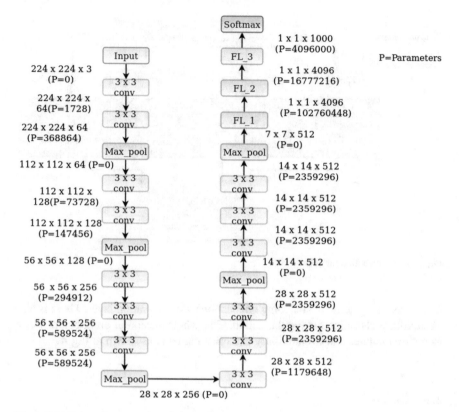

Fig. 2 VGG16 architecture with parameters

made layers not only in depth wise, but also in width wise, by carefully constructing networks that balance depth and width. The inception module shown in Fig. 3, where the module integrates a series of filters, includes a 3×3 filter, 5×5 filter and followed by 1×1 filter with a final concatenation operation. The bottleneck of 1×1 operation reduces computational overhead considerably. This is because, before 3×3 and 5×5 convolution, 1×1 convolution was calculated. The InceptionV3 module has adapted a asymmetric factorization for convolution. It decomposes a 3×3 filter into two, matrix size 1×3 and 3×1 filters respectively, thus reducing the number of learnable parameters from 9 to 6. The module in Fig. 3, takes up an input of $28 \times 28 \times 256$ and undergoes filtering operations with kernels of size 3×3, 5×5 and 1×1 along with max-pooling widthwise and finally a concatenation operation to get an output feature set of $28 \times 28 \times 672$. The computational load is reduced by 28% by factorizing the convolutional kernel of size 5×5 into two 3×3 consecutive kernels.

Apart from the convolutional layer, the network has max-pooling layer and global average pooling layer. Dropouts are successfully implemented for performance optimization. The introduction of dropouts effectively handled the over fitting problems

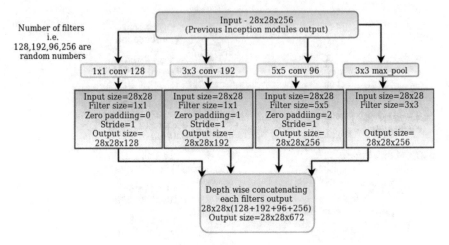

Fig. 3 Naive inception module

in the model, by actively switching connections with activation layer. There is also an auxiliary classifier for regularization. The whole Inception architecture has 8 inception modules and two auxiliary classification layer as shown in Fig. 4.

2.4 ResNet

The key point of ResNet is residual learning and identity mapping by shortcuts using a simple logic depicted in Fig. 5. The skip connection forwards the block inputs as an identity mapping. In normal convolutional networks, going deeper will affect the performance of the network considerably. These networks are prone to the problem of vanishing gradients where the value of gradient become too small during back propagation due to successive multiplication. The residual blocks ensure that back propagation can also be operated without interference [25].

As shown in Fig. 6, residual learning is carried out by feeding the input data directly to the output of single layered or multi-layered convolutional network, by adding them together and further applying the ReLU activation function. Necessary padding is applied on the output convolution network to make it same size as that of the input. Other than tackling vanishing gradients problem, the network encourages feature reuse, thus increasing the feature variance of the outputs. For the 1000 class problem of ImageNet, the network uses 0.85 million learnable parameters.

The ResNet takes up an input of standard size $224 \times 224 \times 3$. As shown in Fig. 6, the architecture has 64 filters of size 7×7 in the first layer. All other filters are of size 3×3 and the number of filers increases gradually up to 512 in the final convolution layer.

Fig. 4 Inception
architecture

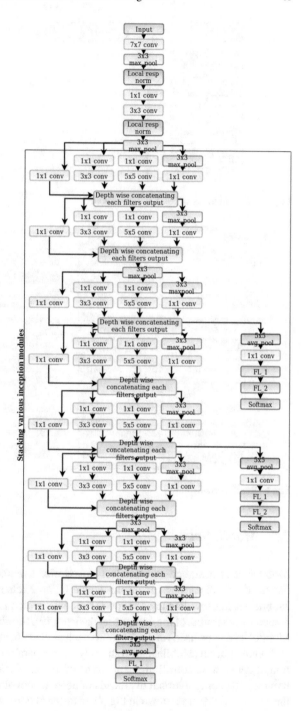

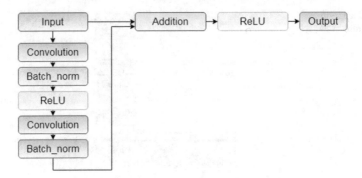

Fig. 5 Residual block of ResNet architecture

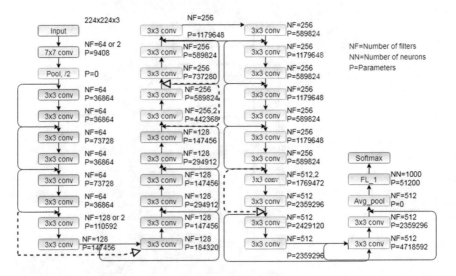

Fig. 6 ResNet architecture

2.5 MobileNet

MobileNet is designed to effectively maximize accuracy, while being careful of the restricted resources for an on-device or an exclusive embedded application. MobileNet has low-latency, low-power models, which is parameterized to meet the resource constraints of the computing devices [19]. Similar to the large scale models, MobileNet can be used for applications such as classification, detection, embeddings and segmentation. MobileNet exclusively uses separable filters, which is a combination of depth-wise convolution and point-wise convolution. It uses 1×1 filters for reducing the computational overheads of normal convolution operation. This makes the network lighter (as shown in Fig. 7), in terms of computational complexity as well as size. The MobileNet- 224 has around 5 million parameters on ImageNet classi-

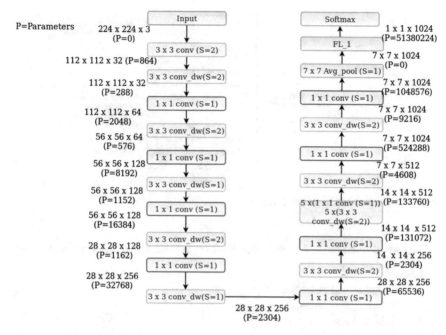

Fig. 7 MobileNet architecture with parameters

fication. MobileNet also takes up an input of size $224 \times 224 \times 3$. The number of filters may vary from 32 to 1024 as in Fig. 7.

2.6 NASNet

The deep neural network has witnessed growth to the next generation by introducing the concept of optimized network. This idea was materialized through the concept of NAS by Google ML group. Their approach was based on reinforcement learning. The parent AI reviews the efficiency of the child AI and makes adjustments to the neural network architecture. Several modifications was done based on number of layers, weights, regularization methods, etc., to improve efficiency of the network. The architecture consists of Controller Recurrent Neural Network (CRNN) and CNN, which is to be trained as depicted in Fig. 8. The NASNet A, B, C version algorithms, choose the best cells by the reinforcement learning method as in [21]. According to Chen et al. [26], reinforced evolutionary algorithm can be implemented for selecting the best candidates. Tournament selection algorithm is implemented to eliminate the worst performing cell. The child fitness function is optimized and reinforcement mutations are performed. This further optimizes the performance of the cell structure.

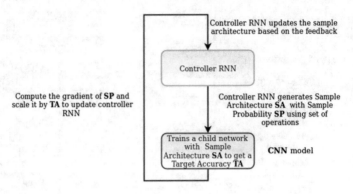

Fig. 8 Controller RNN in NASNet architecture

The set of operational blocks in the NASNet architecture is listed as below:

- Identity
- 1×3 then 3×1 convolution
- 1×7 then 7×1 convolution
- 3×3 dilated convolution
- 3×3 average pooling
- 3×3 max pooling
- 5×5 max pooling
- 7×7 max pooling
- 1×1 convolution
- 3×3 convolution
- 3×3 depthwise-separable—convolution
- 5×5 depthwise-separable—convolution
- 7×7 depthwise-separable—convolution

The NASNet architecture is trained with two types of input images of size 331×331 and 224×224, to get NASNetLarge and NASNet Mobile architectures respectively. There is an extensive growth in number of parameters while moving from NASNetMobile to NASNetLarge. NASNetMobile has 53,26,716 parameters and NASNetLarge has 8,89,49,818 parameters. This makes NASNetMobile more reliable.

Block is the smallest unit in NASNet architecture and cell is the combination of blocks, as shown in Fig. 9. A cell is formed by concatenating different blocks as in Fig. 10. The search space introduced in NASNet is by factorizing the network into cells and further subdividing it into blocks. These cells and blocks are not fixed in number or type, but are optimized for the selected dataset.

Block is an operational module as depicted in Fig. 11. The possible operations of a block include normal convolutions, separable-convolutions, max-pooling, average-pooling, identity mapping etc. The blocks maps two inputs (present and one previous, say H_0 and H_1) to a single output feature map. It takes element wise addition. If the

Fig. 9 Taxonomy of
NASNet architecture

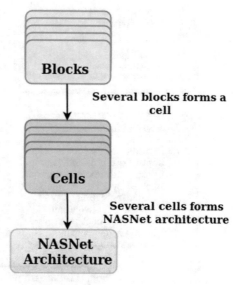

Fig. 10 Formation of a cell
in NASNet architecture

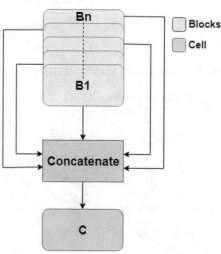

cell takes a block having feature map of size H × W and stride of 1, output will be the
same size as that of feature map. If the stride is 2, the size will be reduced by 2. The
cells are combined in an optimized manner. The network development is focused on
three factors: the cell structure, the number of cells to be stacked (N) and the number
of filters in the first layer (F). Initially N and F are fixed during the search. Later N
and F in the first layer are adjusted to control the depth and width of network. Once
search is finished, models are constructed with different sizes to fit the datasets.

The cells are then connected in an optimized manner to develop into the NAS-
Net architecture. Each cell is connected to two input states, termed as hidden state.

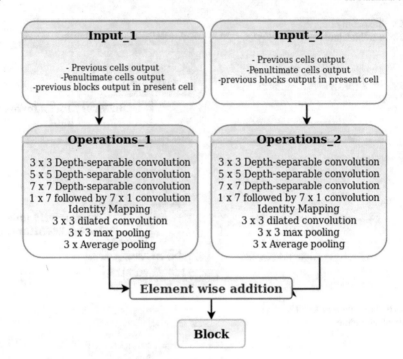

Fig. 11 Formation of a block in NASNet architecture

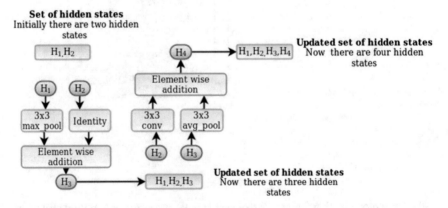

Fig. 12 Formation of hidden states inside a block

A sample of hidden state formation (4 hidden layers) is shown in Fig. 12. The hidden layers are formed through pairwise combinations and updated by concatenation operation.

Hidden layers can undergo a set of convolution and pooling operations. Instead of searching the entire cells, only the best cells are selected in NASNet architecture. This will make the search faster and thus more generalized features could be obtained. The NAS search space has a controller-child structure, where the controller is a normal cell and the child is a reduction cell, as shown in Figs. 13 and 14 respectively. Normal

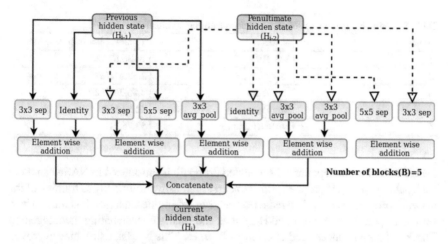

Fig. 13 Normal cell in NASNet architecture

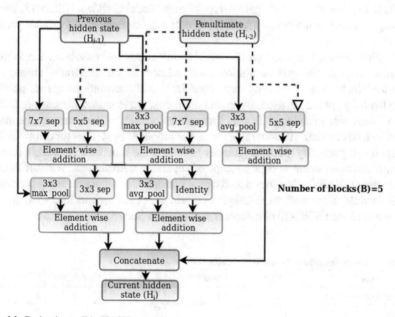

Fig. 14 Reduction cell in NASNet architecture

cell maintains the height and width dimension as same as that of the input feature map where as, the reduction cell reduces these dimensions by two. N number of normal cells are stacked in between the reduction cells to form NASNet architecture. To provide high accuracy NASNetLarge has taken N as 6 while the main concern for NASNetMobile was to run with limited resources.

For both normal cell and reduction cell, the input size of $224 \times 224 \times 3$ is reduced to a size of 7×7 at the output through the selected set of operations using

Table 3 Learnable parameters of standard deep learning architectures

Sl.no	Model	Total number of parameters
1	VGG19	14,36,67,240
2	InceptionV3	2,38,51,784
3	ResNet50	2,56,36,712
4	MobileNet	42,53,864
5	NASNet mobile	53,26,716

5B cells. A new concept named scheduled droppath is introduced in NASNet, where each path in the cell is dropped with a linearly increasing probability as training of the network progresses [21]. The reduction cell Fig. 14 implies that the H_i is formed by a series of operations on H_{i-1} and H_{i-2} cells and finally concatenating them together. The RNN repeats these prediction steps B times. This NASNet structure improved the speed of operation remarkably.

Total number of learnable parameters in each model is shown in Table 3. While coming to newer architectures a scenario on reducing the number of learnable parameters was observed.

Major operations in the above discussed architectures are consolidated in Table 4. It is observed that during the evolution, each architecture has contributed unique features to CNN models. In the early stage, combination of convolution and max-pooling were the only operations used. By giving more concern to spatial dimension, Inception module was introduced. During those times, vanishing gradient was the threat faced by CNN models. An admirable move was proposed by ResNet for resolving this problem using identity mapping. Inorder to reduce the number of learnable parameters, MobileNet came with a concept of separable-convolution. Without human involvement, NASNet has introduced a new optimized architecture using controller RNN module. How well these unique operations in each architecture were able to attain results on UERC-2019 dataset is discussed in Sect. 5.

Table 4 Salient operations of standard architectures

Operations	VGG	Inception	ResNet	MobileNet	NASNet
Normal convolution	Y	Y	Y	Y	Y
Identity			Y		Y
Average pooling			Y	Y	Y
Max pooling	Y	Y			Y
Separable—convolution				Y	Y
Batch normalization				Y	Y
Global average pooling		Y	Y	Y	Y
Factorization		Y			
Grid size reduction		Y			
Auxiliary classification		Y			
Scheduled Drop Path					Y

3 UERC-2019 Dataset Description

As ear recognition is becoming one of the most promising consideration for person identification, researchers has shown a large interest for resolving various problems equipped with ear biometric. In the beginning, researchers concentrated mainly on images of ears in constrained environment, but recent studies are progressing in unconstrained environment. Some of the available datasets for ear recognition are AWE [4], AWEx [27], WebEars [28] etc., Performance analysis of the architectures presented in Sect. 2 has evaluated using UERC-2019 dataset. Open problems emerged from UERC-2017 challenge has motivated to conduct the second series of the challenge (UERC-2019). Apart from 2017, UERC-2019 challenge focuses on the effectiveness of technology on various aspects such as gender and ethnicity [7]. Figure 15 depicts the UERC-2019 dataset, which comprises of 11,500 RGB images from 3,740 subjects. The dataset consists of varieties of realistic images obtained using web spiders, which makes automatic ear identification a difficult task. It includes images of size varying from 15×29 pixels to 473×1022 pixels, which shows the divergent nature of the input data in terms of image resolution as well as visual quality.

Sample images from UERC-2019 dataset are shown in Fig. 16. The dataset is basically branched into two parts such as public dataset and sequestered dataset. The sequestered dataset was provided only to the participants in the challenge, which consist of 500 images from 50 subjects for crosschecking the algorithms developed by the participants. So, we have chosen publically available UERC-2019 dataset for our analysis. Public dataset was split into two divisions as training and testing datasets. The images for training dataset was entirely taken from AWEx [6, 29]. The test data from public dataset contains 9,500 images from 3,540 subjects, taken from AWEx and UERC-2017 dataset. Annotations were not provided for test data as well as some subjects in the train data. The train data in UERC-2019 has 166 subjects, in which annotations were provided till 150 subjects with equal distribution of images (10 images per subject), remaining subjects appeared without annotations as well as

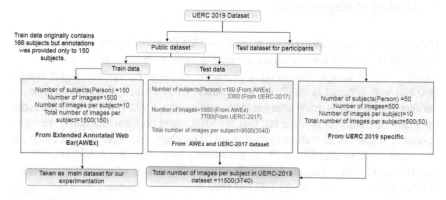

Fig. 15 UERC-2019 dataset

Fig. 16 Sample of images from UERC-2019 dataset

uneven distribution of images. Since the number of images in the training dataset is small, various augmentation techniques has been used inorder to increase the size of the data.

4 Design of Experiment

Data for all the experiments were taken from UERC-2019 dataset. Experiment was evaluated on 5 architectures which are having unique characteristics and are commonly used for image classification problems in Keras framework. Even though NASNet was released in 2017, we were unable to find much of its applications in various domains. This made us eager to learn more about NASNet architecture. From the study on Sect. 2, it is observed that NASNet could perform efficiently on small dataset (CIFAR-10) as well as large dataset (ImageNet). Since the dataset taken for our experiment is also small, the intention was to explore the NASNet architecture on UERC-2019 dataset. As NASNetMobile has lesser number of parameters, computations were possible on GPU as well as in CPU. Pre-processing techniques plays a vital role in image classification tasks such as upgrading features of images,

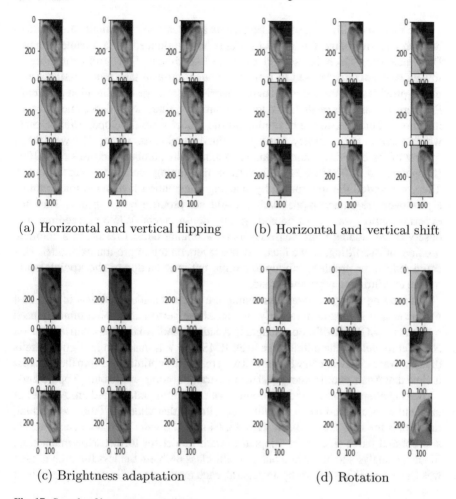

(a) Horizontal and vertical flipping   (b) Horizontal and vertical shift

(c) Brightness adaptation   (d) Rotation

Fig. 17 Sample of image augmentation

increasing the number of images etc. Initially, experiments were carried out on various pre-processing techniques such as data augmentation and data normalization. We have tried commonly used image augmentation techniques like shifting, flipping, brightness adaptation, rotation, zooming etc. Sample results obtained from each augmentation technique is shown in Fig. 17.

Environment Setup

- **Processor**: Intel@ Core™ i5-7200U CPU
- **Platform**: Google Colaboratory (GPU support)
- **RAM**: 16 GB (CPU), 12.72(GPU)
- **Disk space**: 358.27GB (GPU)
- **Python version**: 2.7
- **Python framework**: Keras framework (2.2.4) with Tensorflow (1.13.1) backend.

The memory utilization of traditional data augmentation techniques are found to be more. To overcome this problem, Keras has introduced a class named Image-DataGenerator, which is used all over the experiment. Flow_from_directory in ImageDataGenerator is used to make the image retrieval task much easier. Keras model provides pre-trained weights of ImageNet for image classification problems. Performance analysis of all the models without pre-trained network became a curious intention. Therefore, in the beginning all the model were developed from scratch, which result in low accuracy. Later pre-trained weights from ImageNet were taken for rest of the experimentation. Inorder to reduce the number of classes from 1000 (ImageNet) to 150 (UERC-2019), various fine-tuning techniques were adopted. Then, we started the fine-tuning by stacking convolution layers and max-pooling layers over pre-trained model which, could not produce better results. Then the experimentation was based on adding only dense layers. While proceeding, it is observed that adding more dense layers on a smaller dataset results in a common problem of overfitting, so we fixed 2 dense layers on top of pre-trained model. This could reduce the problem only up to a limit, but introduction of checkpoints could reduce overfitting to a greater extend.

On moving to hyper parameter tuning the main consideration was to find best optimizer and it's learning rate. Evaluation of the performance of commonly used optimizers (SGD, RMSprop, Adagrad, Adadelta and Adam) was carried out on NASNet model. At the initial stage itself RMSprop was unable to give good results thus the evaluation was carry forwarded with rest of the optimizers. From the Fig. 18 it is clear that Adam could achieve best result in training compared to other 3 optimizers.

Next challenge was to find out optimal value learning rate for Adam. A series of trial and error method came up with a conclusion that choosing 0.001 as learning rate yields the best result. Batch sizes 32, 64 and 128 were tried with each models and optimal batch sizes were adopted to each model for further experimentation. Since the analysis to be carried out on multi-class problem the loss function chosen was Categorical Cross-Entropy. In general, each model was build using a pre-trained

Fig. 18 Comparison of performance of optimizers on the dataset for NASNet model

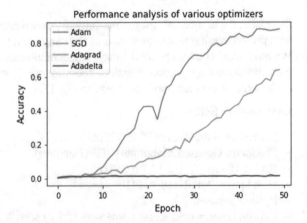

model stacked with 2 dense layers, Categorical Cross-Entropy as loss function and Adam as optimizer with a learning rate of 0.001. Finally each model were deployed for testing.

5 Results and Discussion

The constructed model was deployed for evaluation. Primary layer feature extraction on each model is shown in Fig. 19.

Figure 20 shows representation of feature extraction in intermediate blocks on different models. As we could see various models are learning features in a different manner. Like this, on each blocks the features learned are updated.

The results exhibited same behaviour even scaling to higher number of epochs. Therefore, we kept a baseline of 100 epochs which is reliable in CPU as well. Based on the fair experiments mentioned above, we observed that NASNet could perform slightly better while comparing to other models for this particular experimental setup. The epoch which gave highest testing accuracy throughout the experiment has given

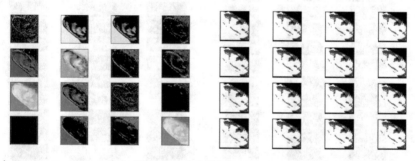

(a) Feature map from first CNN layer of VGG19 (b) Feature map from first CNN layer of ResNet

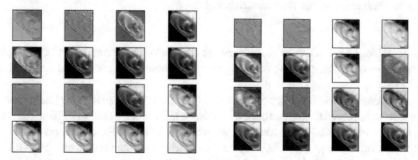

(c) Feature map from first CNN layer of MobileNet (d) Feature map from first CNN layer of NASNet

Fig. 19 Feature map on each model

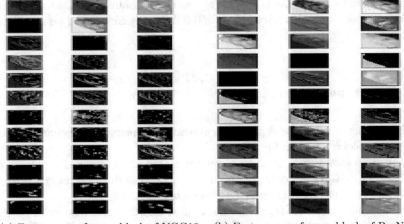

(a) Feature map from a block of VGG19   (b) Feature map from a block of ResNet

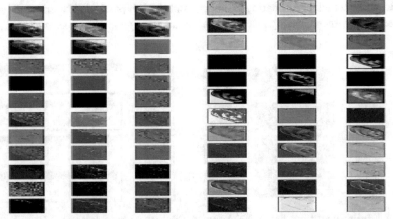

(c) Feature map from a block of MobileNet   (d) Feature map from a block of NASNet

Fig. 20 Feature map from a block on each model

in the table. Other criterion considered for the performance analysis of developed model was computation time, batch size and total number of parameters depicted in Table 5.

InceptionV3 was showing poor accuracy (less than 5%) from training itself, therefore we were forced to stop further evaluation on this model. Since VGG16 also could not yield better result while training, we went for VGG19. Pictorial representation of final models are given in Fig. 21. As we could see all the models except VGG19 were performing well during training period, while there was a drastic drop in test accuracy.

Table 5 Comparison of models in terms of computational time, batch size and total no. of parameters

Model	Total no. of epochs	Computation time (s)	Batch size	Total no. of parameters
VGG19	100	19	128	3,08,40,106
Resnet50	100	13	32	3,44,48,086
MobileNet	100	11	128	7,50,45,398
NASNet	100	16	128	2,37,87,734

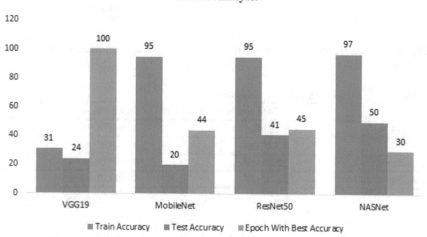

Result Analysis

Train Accuracy   Test Accuracy   Epoch With Best Accuracy

Fig. 21 Performance analysis of standard CNN architectures based on classification accuracy

Table 6 The performance metrics using NASNet model

Accuracy	Precision	Recall	F1-Score
50.4%	47.3%	47.1%	53.4%

Diagrammatic representation of epoch with high accuracy is depicted in Fig. 22. From the figure it is clear that NASNet could achieve it's best accuracy on an average of 30 epoch but the same could be achieve by ResNet, MobileNet and VGG19 on 45, 44 and 100 epochs respectively.

The optimized search architecture of the NASNet model has made it a suitable candidate for unconstrained ear recognition over the other architectures considered here. The consistency of the performance was further evaluated through the performance measures such as accuracy, precision, recall and f1-score as an average of 150 classes. The performance measures obtained for ear recognition using NASNet model is tabulated in Table 6.

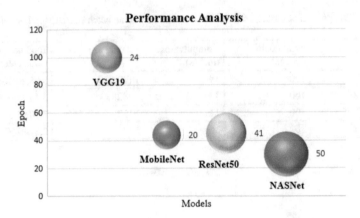

Fig. 22 Performance analysis of standard architectures used for ear recognition based on number of epochs

Table 7 GPU memory utilization of various models

Model	RAM size out of 12.72 GB	Disk space out of 358.27 GB
VGG19	2.51 GB	28.04 GB
ResNet	3.35 GB	31.16 GB
MobileNet	2.82 GB	40.10 GB
NASNet	5.22 GB	30.82 GB

The Table 7 shows memory utilization of each model on GPU in our constrained environmental setup. It is observed that all the models have used limited amount of memory as well as disk space.

6 Conclusions and Future Directions

Applications of deep neural networks are growing exponentially and continues to be one of the major verticals of AI. The prime objective of this chapter was to analyse the performance of unique DL architectures, especially on trivial problems of biometric classifications using unconstrained ear dataset. We explored the major features of the selected state of the art networks. Along with that, the possibilities of the new paradigm of optimized deep neural network was also analysed. From the study we could observe that many of the selected architectures could perform well in training but during evaluation with small test data provided, there was a drop in performance. In this constrained environmental setup we could observe that NASNet was giving moderately better results. This opens up a new door for optimized neural networks in image classification. By incorporating datasets of larger size, better models can be generated. In that case, the results could be further enhanced. Further enhancements has to be done in order to improve the performance of the system using various environmental setup and also concentrating on each model separately. Thus, in future, ear as a bio metric identification method can be introduced by fine tuning the training models. A person can be identified by applying the test image from any environment.

References

1. LeCun, Yann, Yoshua Bengio, and Geoffrey Hinton. 2015. Deep learning. Nature 521.7553: 436.
2. Hatcher, William Grant, and Wei Yu. 2018. A survey of deep learning: platforms, applications and emerging research trends. *IEEE Access* 6: 24411–24432.
3. Sundararajan, Kalaivani, and Damon L. Woodard. (2018). Deep learning for biometrics: A survey. *ACM Computing Surveys (CSUR)* 51 (3): 65.
4. Kumar, Ajay, and Wu Chenye. 2012. Automated human identification using ear imaging. *Pattern Recognition* 45 (3): 956–968.
5. Nejati, Hossein, Li Zhang, Terence Sim, Elisa Martinez-Marroquin, and Guo Dong. (2012). Wonder ears: Identification of identical twins from ear images. In *Proceedings of the 21st International Conference on Pattern Recognition (ICPR2012)*, 1201–1204. IEEE.
6. Emersic, Ziga, Dejan Stepec, Vitomir Struc, Peter Peer, Anjith George, Adii Ahmad, Elshibani Omar et al. 2017. The unconstrained ear recognition challenge. In *2017 IEEE International Joint Conference on Biometrics (IJCB)*, 715–724. IEEE.
7. Emersic, Ziga, B. S. Harish, Weronika Gutfeter, Jalil Nourmohammadi Khiarak, Andrzej Pacut, Earnest Hansley, Mauricio Pamplona Segundo et al. 2019. The unconstrained ear recognition challenge 2019-arxiv version with appendix. arXiv:1903.04143.
8. Alom, Md Zahangir, et al. 2018. The history began from AlexNet: A comprehensive survey on deep learning approaches. arXiv:1803.01164.
9. Unnikrishnan, Anju, V. Sowmya, and K. P. Soman. 2019. Deep learning architectures for land cover classification using red and near-infrared satellite images. *Multimedia Tools and Applications* 1–16.
10. LeCun, Yann, Léon Bottou, Yoshua Bengio, and Patrick Haffner. 1998. Gradient-based learning applied to document recognition. *Proceedings of the IEEE* 86 (11): 2278–2324.
11. Krizhevsky, Alex, Ilya Sutskever, and Geoffrey E. Hinton. 2012. Imagenet classification with deep convolutional neural networks. In *Advances in Neural Information Processing Systems*, 1097–1105.
12. Zeiler, Matthew D., and Rob Fergus. 2014. Visualizing and understanding convolutional networks. In *European Conference on Computer Vision*, 818–833. Springer, Cham.
13. Lin, Min, Qiang Chen, and Shuicheng Yan. 2013. Network in network. arXiv:1312.4400.
14. Simonyan, Karen, and Andrew Zisserman. 2014. Very deep convolutional networks for large-scale image recognition. arXiv:1409.1556.
15. Szegedy, Christian, Wei Liu, Yangqing Jia, Pierre Sermanet, Scott Reed, Dragomir Anguelov, Dumitru Erhan, Vincent Vanhoucke, and Andrew Rabinovich. 2015. Going deeper with convolutions. In *Proceedings of the IEEE Conference on Computer Vision and Pattern Recognition*, 1–9.
16. He, Kaiming, Xiangyu Zhang, Shaoqing Ren, and Jian Sun. 2016. Deep residual learning for image recognition. In *Proceedings of the IEEE Conference on Computer Vision and Pattern Recognition*, 770–778.
17. Larsson, Gustav, Michael Maire, and Gregory Shakhnarovich. 2016. Fractalnet: Ultra-deep neural networks without residuals. arXiv:1605.07648.

18. Huang, Gao, Zhuang Liu, Laurens Van Der Maaten, and Kilian Q. Weinberger. 2017. Densely connected convolutional networks. In *Proceedings of the IEEE Conference on Computer Vision and Pattern Recognition*, 4700–4708.
19. Howard, Andrew G., Menglong Zhu, Bo Chen, Dmitry Kalenichenko, Weijun Wang, Tobias Weyand, Marco Andreetto, and Hartwig Adam. 2017. Mobilenets: Efficient convolutional neural networks for mobile vision applications. arXiv:1704.04861.
20. Chollet, François. 2017. Xception: Deep learning with depthwise separable-convolutions. In *Proceedings of the IEEE Conference on Computer Vision and Pattern Recognition*, 1251–1258.
21. Zoph, Barret, Vijay Vasudevan, Jonathon Shlens, and Quoc V. Le. 2018. Learning transferable architectures for scalable image recognition. In *Proceedings of the IEEE Conference on Computer Vision and Pattern Recognition* 8697–8710.
22. Kurup, R. Vimal, V. Sowmya, and K. P. Soman. 2019. Effect of data pre-processing on brain tumor classification using capsulenet. In *International Conference on Intelligent Computing and Communication Technologies* 110–119. Singapore:Springer.
23. Sabour, Sara, Nicholas Frosst, and Geoffrey E. Hinton. 2017. Dynamic routing between capsules. In *Advances in Neural Information Processing Systems* 3856–3866.
24. Szegedy, Christian, Vincent Vanhoucke, Sergey Ioffe, Jon Shlens, and Zbigniew Wojna. 2016. Rethinking the inception architecture for computer vision. In *Proceedings of the IEEE Conference on Computer Vision and Pattern Recognition* 2818–2826.
25. K. He, X. Zhang, S. Ren, J. Sun. 2016. Deep residual learning for image recognition. In: *2016 IEEE Conference on Computer Vision and Pattern Recognition (CVPR)* 770–778.
26. Chen, Yukang, Qian Zhang, Chang Huang, Mu Lisen, Gaofeng Meng, and Xinggang Wang. 2018. *Reinforced evolutionary neural architecture search*. CoRR.
27. Emersic, Žiga, Blaž Meden, Peter Peer, and Vitomir Štruc. 2018. Evaluation and analysis of ear recognition models: Performance, complexity and resource requirements. *Neural Computing and Applications* 1–16.
28. Zhang, Yi, and Zhichun Mu. 2017. Ear detection under uncontrolled conditions with multiple scale faster region-based convolutional neural networks. *Symmetry* 9 (4): 53.
29. Emersic, Z., V. Struc, and P. Peer. 2017. Ear recognition: More than a survey. *Neurocomputing* 255: 26–39.

Optimization of Performance Parameter for Vehicular Ad-hoc NETwork (VANET) Using Swarm Intelligence

Biswa Ranjan Senapati and Pabitra Mohan Khilar

Abstract Vehicular Ad-hoc NETwork (VANET) is a subcategory of Mobile Ad-hoc NETwork (MANET) which is one of the popular emerging research areas. On one side increase in the number of vehicles, and on the other side due to the presence of communication unit i.e. On Board Unit (OBU) in vehicles helped in the creation of a new network called as VANET. VANET is becoming so popular that it is widely used for different applications that may be safety or non-safety applications. An effective routing protocol is required for the efficient use of VANET for different applications which can optimize different generic parameters of VANET like end-to-end delay, number of hops, etc. Altough multiple paths exist between a source and a destination but the routing protocols evaluate a single path for the transmission of information packets based on parameters like shortest distance towards the destination, density of vehicles, number of hops, etc. Different swarm intelligence techniques like Ant Colony Optimization (ACO), Particle Swarm Optimization (PSO), etc. can be used in VANET to optimize the parameters used in the routing protocol. In this chapter, Ant Colony Optimization (ACO) is used to establish multiple routes between nodes which is vital in the network where network connectivity is random and is frequently changing. This also helps in the parallel transmission of packets in multi-path which reduces end-to-end delay. For this proposed work, a variation of end-to-end delay with respect to transmission distance and a number of vehicles is compared with the existing VANET routing protocol i.e. Geographic Source Routing (GSR), Anchor based Street Traffic Aware Routing (A-STAR). The comparison shows that the proposed work performs better as compared to GSR and A-STAR. To implement multi-path routes MongoDB—an open source distributed database is used and the corresponding operation on the database is implemented using node-red. Selection of effective route and intermediate hop for the transmission of emergency information is essential. The optimized route in VANET not only saves the overall transmission time but

B. Ranjan Senapati (✉) · P. Mohan Khilar
Department of Computer Science and Engineering, National Institute of Technology,
Rourkela, India
e-mail: biswa.rnjn@gmail.com

P. Mohan Khilar
e-mail: pmkhilar@nitrkl.ac.in

© Springer Nature Switzerland AG 2020
M. Rout et al. (eds.), *Nature Inspired Computing for Data Science*,
Studies in Computational Intelligence 871,
https://doi.org/10.1007/978-3-030-33820-6_4

also reduces the end-to-end delay. For the optimization of route and number of hops in VANET, another swarm intelligence technique i.e. Particle Swarm Intelligence (PSO) can be used through which the transmission of emergency information with optimum delay is proposed.

Keywords VANET · MANET · OBU · RSU · ACO · PSO

1 Introduction

Vehicular Ad-hoc NETwork (VANET) is the fastest-growing research area which is used for various applications. To avoid time consumption during transportation and to lead a comfort and luxurious life, the number of vehicle users are increasing day by day. Figure 1 shows the statistics for the number of vehicles manufactured per year in India [1].

Nowadays, vehicles are not used only as the traditional carrier. Due to the presence of various components such as different sensors, communication unit such as OBUs, etc make the vehicle a smart and intelligent carrier. Figure 2 shows various components present in the modern vehicle that makes the vehicle smart and intelligent [2]. Due to the advancement of technology different sensors are developed for various applications at affordable costs. Das et al. surveyed the design architecture, principles of virtual sensor networks [3]. Thus, on one hand, increase in the number of vehicles used in the road network, and on the other hand presence of communication unit in the vehicles helped the researcher to create a new network which is the subset of Mobile Ad-hoc NETwork (MANET) called as Vehicular Ad-hoc NETwork (VANET) [4, 5]. VANET consists of two components.

Fig. 1 Number of vehicles manufactured per year in India

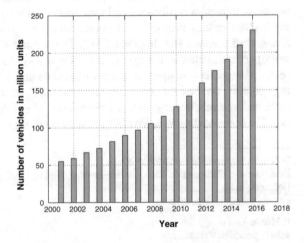

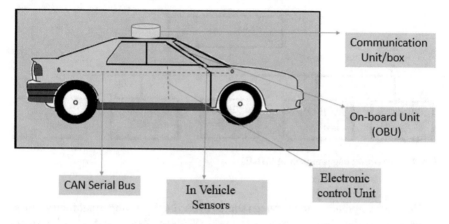

Fig. 2 Smart and intelligent vehicle

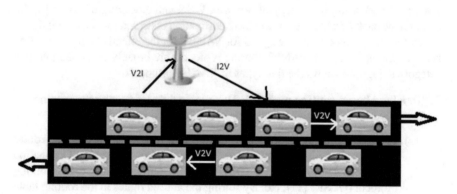

Fig. 3 Types of communication

1. **Vehicles**: These are the mobile nodes of VANET. Vehicles are the vital components of VANET because of the presence of communication unit OBU.
2. **Road Side Units (RSUs)**: These are the static nodes of VANET having high communication range as compared to vehicles.

Based on the two components of VANET, communication in VANET is classified into two categories. These are as follows.

1. Vehicle To Vehicle (V2V)
2. Vehicle To Infrastructure (V2I) or (I2V)

Figure 3 shows the types of communication in VANET. For V2V communication, the receiving vehicle must be within the communication range of the transmitting vehicle. Similarly, for V2I communication, the receiving vehicle or RSU must be within the communication range of transmitting RSU or vehicle respectively.

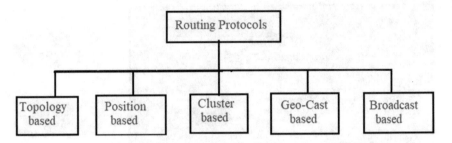

Fig. 4 Types of routing protocols in VANET

Effective communication between source to destination is important for a wireless network. It may be a Wireless Sensor Network (WSN) or Vehicular Ad-hoc NETwork (VANET). Sarkar et al. proposed an architecture for data management in wireless sensor network using the energy of nodes in WSN and clustering of nodes [6]. To transmit the packet from the source node to destination node, routing plays a vital role. Several routing protocols are designed for VANET for the effective communication between the nodes [7–9]. VANET routing protocols are broadly classified into five categories. Figure 4 shows the five types of VANET protocols.

1. **Topology based routing protocol**: This protocol depends upon the topology of the network. This is again broadly classified into two categories.

   (a) **Proactive routing protocol**: In this type of protocols, all possible routes between all nodes are maintained in the routing table. E.g. Destination Sequenced Distance Vector (DSDV) [10], Optimized Link State Routing protocol (OLSR) [11], etc. By storing the unused routes in the routing table increases the network load and also increases the bandwidth consumption. This reduces the performance of the network.

   (b) **Reactive routing protocol**: This protocol removes the limitations of the proactive routing protocol. In reactive protocols routes between the nodes are discovered on demand. E.g. Dynamic Source Routing (DSR) [12], Ad-hoc On-Demand Distance Vector (AODV) [13], Temporally Ordered Routing Algorithm (TORA) [14], etc. Khiavi et al. compared the performance of different network parameters for the routing protocols AODV, DSDV, DSR, TORA [15]. Figure 5 shows the different types of topology-based routing protocols with example.

2. **Position based routing protocol**: In this routing protocol, the next hop to transmit the packet is determined by using geographic location information. By using street maps, GPS service, GIS service the location of next hop is determined. Examples of position based routing protocols are Geographic Source Routing (GSR) [16], Greedy Perimeter Stateless Routing (GPSR) [17], Connectivity Aware Routing Protocols (CARP) [18], Anchor-Based Street and Traffic Aware Routing (A-

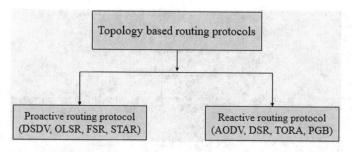

Fig. 5 Types of Topology based routing protocols in VANET

STAR) [19] and so on. Figure 6 shows the taxonomy of position based routing protocols used in VANET.

3. **Cluster based routing protocols**: To make a larger network scalable, cluster based routing protocols are designed. In this network, a cluster is formed by a set of nodes where all the nodes are present within the communication range of a special node called a cluster head. Example of cluster based routing protocols is Cluster Based Routing (CBR) [20], Cluster Based Location Routing (CBLR) [21], Hierarchical Cluster Based routing (HCB) [22] and so on.

4. **Geo cast based routing protocols**: This type of protocols are used to transmit the message to a particular geographical region by multicasting service by using either flooding techniques or non-flooding techniques. Example of Geo Cast based routing protocols are Inter-Vehicle Geocast (IVG) [23], Distributed Robust Geocast (DRG) [24], Dynamic Time Stable Geocast (DTSG) [25] and so on.

5. **Broadcast based routing**: To share emergency information such as road traffic congestion, road accident information messages are broadcasted for the vehicle to vehicle (V2V) communications using broadcast based routing protocols. But if in the network, the density of vehicles is high then this type of protocol creates network congestion, high bandwidth consumptions, etc. Examples of broadcast based routing protocols are BROADCOMM [26], Urban Multihop Broadcast protocol (UMB) [27], Edge-Aware Epidemic Protocol (EAEP) [28] and so on.

Due to the availability of a large number of sensors at affordable cost, an increase in the number of vehicles with a communication unit, VANET can be used for various applications. The types of applications in VANET is broadly classified into four categories. These are safety-oriented applications, commercial applications, convenience applications, and productive applications. Figure 7 shows the types of applications for VANET.

The main objective of VANET for using different applications is to provide users safety information, commercial information, and also make the transport for the users more enjoyable and entertain-able [29]. Safety oriented applications include different services of Intelligent Transportation System (ITS) [30], real time traffic information transmission [31], information transmission of road accident at a location etc [32]. Commercial applications of VANET include the efficient communication

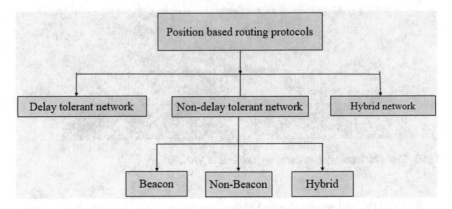

Fig. 6 Taxonomy of Position based routing protocols in VANET

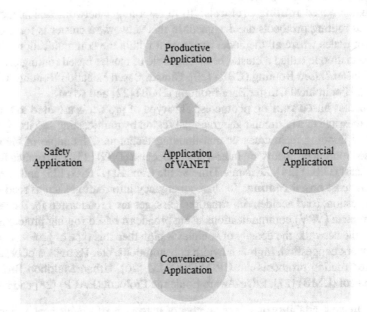

Fig. 7 Different types of application for VANET

between buyer and sellers for the purchase of products [33, 34], sharing of Wifi using VANET [35] etc. Convenience Applications in VANET consists of automatic toll tax collection [36–38], diversion of route during road traffic congestion or due to road accident [39, 40], determination of available parking slot in a mall [41] etc. Productive application in VANET includes monitoring of environmental parameters such as temperature, humidity, wind, rainfall, etc. and transmission of critical information to the vehicle user using VANET [42], saving of fuel and time by the use of automatic toll tax collection [43] etc.

Since VANET is used for several applications, so optimization of different parameters like network delay, signal loss, signal fading, throughput, packet delivery fraction, etc. are essential. The major contribution of this chapter is divided into two parts. In the first part, an emergency message is classified into different category and priority is assigned to each category. For the transmission of an emergency message through VANET PSO is proposed. In the second part, multiple routes between source and destination are established using the concept of ACO. To keep the uniform density of vehicles in the multi-path MongoDB database is used. For this proposed idea, end-to-end delay versus distance and end-to-end delay versus the number of vehicles is compared with the existing VANET routing protocol i.e. GSR and A-STAR. The rest of the chapter is organized as follows. Section 2 discusses the current literature on the application of different swarm intelligence techniques in VANET. System model which comprises of different notations and their meaning, different assumptions, network, and communication model is mentioned in Sect. 3. Section 4 presented the proposed work. Section 5 discusses the result which consists of simulation set up and simulation result. Finally, conclusion and future scope is mentioned in Sect. 6.

2 Related Work

Swarm Intelligence (SI) is nature inspired meta-heuristics optimization techniques [44]. SI is considered as the subset of artificial intelligence (AI) which is applicable for the decentralized and self-organized system [45]. Currently, in various research areas of cloud computing which involves load balancing, work flow scheduling, cost management, etc, uses different swarm intelligence techniques for the optimization of different resources [46]. Broadly the spectrum of swarm intelligence consists of Particle Swarm Optimization (PSO) [47], Ant Colony Optimization (ACO) [48], Artificial Bee Colony (ABC) optimization [49], Bacteria Foraging Optimization (BFO) [50] etc. VANET also uses these broad spectrum of Swarm Intelligence for different application to optimize the performance metric.

Rana et al. proposed a multi-path ACO routing to determine multiple routes between the nodes [51]. This routing is proposed by considering different characteristics of the motion of a vehicle such as the density of the vehicle, the movement pattern of the vehicle, speed of the vehicle, etc. This protocol makes the vehicular network more scalable. Sahoo et al. proposed a cluster based routing protocol using Ant Colony Optimization to make the vehicular network scalable [52]. In this protocol, clusters are created considering the parameters such as the direction of motion of vehicles, speed, and position of the vehicles. For every cluster, cluster head is created by the real-time updated position and by using the trust value of the vehicles present in the cluster.

M. Shoaib and W.C. Song proposed a multi-objective data aggregation technique using Particle Swarm Intelligence (PSO) for VANET [53]. J. Toutouh and E. Alba proposed a parallel PSO (pPSO) and applied to one VANET routing protocol called Ad-hoc On-demand Distance Vector (AODV) routing protocol which showed the better result in different generic QoS metric [54]. Zukarnain et al. proposed a method to enhance the performance of Optimization Link State Routing Protocol (OLSR) by PSO [55]. Due to frequent link disruption and random speed of the vehicle, position based routing protocol is preferred in VANET. O. Karwartya and S. Kumar proposed a mechanism of geocasting among the set of vehicles within a region by using PSO [56]. For solving linear and nonlinear problem PSO is globally accepted. Das et al. proposed a PSO based evolutionary multilayer perceptron for the classification of a task in data mining [57].

The Quality of Service (QoS) for the routing protocol of VANET has to be improved after the addition of multimedia service to VANET. Fekair et al. proposed a method for the improvement of QoS of the routing protocols using cluster based Artificial Bee Colony optimization technique [58]. The QoS for the multicast routing is considered as the NP complete problem. The performance of the swarm intelligence technique is better than the classical problem for the above NP complete problem. Zhang et al. proposed an algorithm called micro artificial bee algorithm (MABC) to improve the QoS for the multicast routing in VANET [59]. The connectivity between the nodes plays an important role in the reliable transmission of information in VANET. Connectivity between all the nodes in a network is effectively represented by minimum spanning tree (MST). The classical algorithm in graph theory could determine only one MST. But swarm intelligence techniques provide several alternatives for the MST. Zhang et al. proposed a binary artificial bee colony algorithm to determine several alternatives MST which is helpful for the reliable transmission of information in VANET [60]. Connectivity between the nodes i.e. either between the vehicles or RSUs are not static. The connectivity is changing frequently due to the random speed of the vehicles. Baskaran et al. proposed a dynamic pattern of connectivity for VANET using Artificial Bee Colony approach [61]. The literature study shows that different swarm intelligence techniques are used for enhancing the performance of VANET.

3   System Model

This section of the chapter mentioned different notations and their meaning used for the proposed work. This section also described different assumptions, network model, and communication model used in VANET.

3.1   Different Notations and Their Meaning

Different notations and their meanings used in the proposed work and also in the network model and communication model are represented in Table 1.

Table 1 List of Notations and their meanings

Notations	Meanings
Veh	Vehicle
RSU	Road side unit
OBU	On board unit
Jn	Junction
PTH_{VAL}	Path value
T	Time period
d	Distance
V	Speed
PSO	Particle swarm optimization
ACO	Ant colony optimization
TTL	Time to leave
LCS	Link connectivity stability
CR	Communication range

3.2 Assumption

1. All the RSUs have fixed infrastructures i.e. they are static in nature.
2. The mobile vehicles and RSUs participate in the communication process.
3. Communication between the nodes in the VANET is either unicast if the source and destination are within the communication range of each other or multicast if the nodes(source and destination) are not present within the communication range.
4. The connectivity links between the nodes are symmetric.
5. The processing and storage capacity of RSUs is more than the mobile nodes i.e. vehicles.
6. The position and speed of the vehicle are determined by GPS.
7. All the vehicles have information about the network channel status and network topology.
8. For ACO, the motion of the vehicles is assumed to be as the motion of ants.

3.3 Network Model

The road framework is modeled as a graph G. In this graph, the junctions Jn_1, Jn_2, ..., Jn_n is considered as the set of vertex and the connected roads between the junctions are considered as the set of edges $E_1, E_2, ..., E_n$. Vehicular networks consist of a set of nodes which are either mobile (vehicles) or static (RSUs). All the nodes in the network can be the sender, receiver, forwarder, or router. The topology in VANET is random, dynamic, and frequently changing due to the random motion of the vehicles.

3.4   Communication Model

The communication unit of the vehicle is the On Board Unit (OBU). The OBU of one vehicle can communicate with the OBU of other vehicles or with RSU if the nodes are within the communication range of each other. If the source and destination node are not present within the communication range, then multi-hop communication takes place.Efficient algorithm must be used to make the communication between the source and destination cost-effective [62]. For data communication between the nodes, VANET uses Dedicated Short Range Communication (DSRC) standard which follows the IEEE 802.1 standard [63, 64]. Currently, Wireless Access in Vehicular Environment (WAVE) standard based on IEEE 802.1p is also used for the communication between the nodes in VANET [65].

4   Proposed Work

In the proposed work two different swarm intelligence techniques are used for the different application of the vehicular network. First, PSO is proposed to transmit emergency information through VANET for different safety applications. Second, to establish multiple routes between the source and destination in VANET, ACO is used.

4.1   Emergency Information Transmission Through VANET Using PSO

PSO is a swarm intelligence technique whose algorithm is based on the motion of a group of birds in the sky or the motion of a group of fishes in the water. In PSO, the optimal position of the particle is obtained by adjusting the velocity of the individual particle. The general equation for PSO can be represented as follows.

$$v_{next} = v_{Current} + \phi_1(x_{localbest} - x_{current}) + \phi_2(x_{globalbest} - x_{current}) \tag{1}$$

$$x_{next} = x_{current} + v_{next} \times \Delta t \tag{2}$$

The overall operations of PSO is represented by the Fig. 8.

The objective of VANET is to provide a safety application. For safety applications, the transmission of safety information message plays a vital role. Delay in the transmission of safety information or emergency information can cause huge loss of life, health, and wealth. The normal approach which is followed for the transmission of safety or emergency information is the broadcasting of information to all the neighboring vehicles. But this transmission has the following limitations.

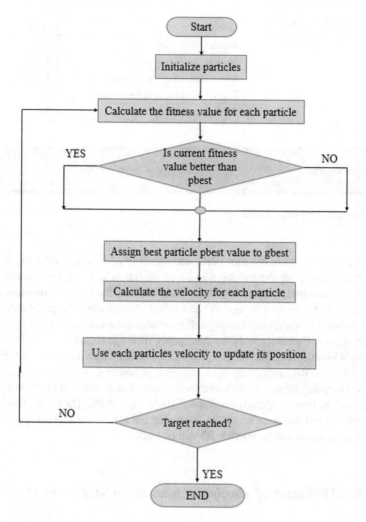

Fig. 8 Overall operations of PSO

1. Apart from the source node, if the forwarder node broadcast the message, then it leads to wastage of bandwidth.
2. Simultaneous broadcast of the message may lead to a hidden node problem.
3. Simultaneous broadcast of the message may not be able to reach a large number of vehicles to take the necessary actions.

The proposed paper classified the emergency message into different categories and assigns priority to different emergency message. Table 4 shows the different category of the emergency message and the corresponding priority of the message (Table 2).

Table 2 Classification of emergency message and its priority

Category	Priority
Break failure warning	1
Vehicle accident warning	2
Intersection of roads warning	3
Fire location information warning	4
Road jam warning	5

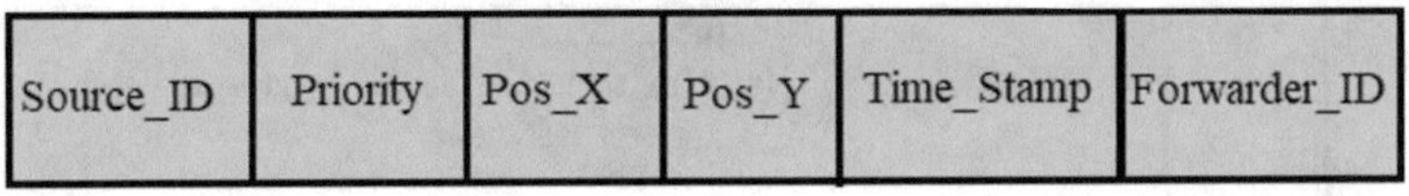

Source_ID	Priority	Pos_X	Pos_Y	Time_Stamp	Forwarder_ID

Fig. 9 Emergency message format

If the forwarder node has more than one category of message, then the decision of the transmission of the message is based on the priority of the message. The emergency message having lesser priority is transmitted first i.e. the contention time is less for the lower priority message. When a vehicle determines a danger, it determines the priority of the emergency message. It generates the emergency message, having a certain format. Figure 9 shows the emergency message format.

In Fig. 9 Source_ID is the vehicle_ID which is unique for all the vehicles. Priority is the priority of the emergency message. Pos_X and Pos_Y is the position of the location where danger has been determined. Time_Stamp refers to the time at which the danger has been detected. Forwarder_ID is the vehicle_ID of the forwarding node which is not the source vehicle. Thus, for the source vehicle transmitting the emergency message the forwarder_ID will be null.

4.2 Establishment of Multiple Routes in VANET by ACO

For the transmission of information from the source to destination, the role of routing protocol is important. For effective routing, the establishment of multiple routes between the source and destination is essential. This is because the connectivity between the nodes in VANET is dynamic and is frequently changing. The link connectivity (LC) between the two nodes depends on two factors. The first factor is to determine the location and speed of the vehicle to establish the link. This determination is done by GPS. The second factor is the establishment of the link. The establishment of a link between the nodes depends upon the distance between the nodes, speed, and communication range. Let Veh_1 and Veh_2 are the two vehicles. Let the position of veh_1 is (x_1, y_1) and the position of Veh_2 is (x_2, y_2). The distance between the vehicles is determined by the following equation.

$$d = \sqrt{(x_2 - x_1)^2 + (y_2 - y_1)^2} \tag{3}$$

Let CR is the communication range. If d < CR, then Veh_1 and Veh_2 are connected. Establishment of multiple connections is possible if the number of vehicles in a particular region will increase. To increase the number of vehicles in a particular region, ACO is used. ACO is an SI technique, which is based on the behavior of ants which have certain characteristics like limited individual capability and visual and auditory communications. Ants are able to connect together because of a chemical substance called pheromones. Figure 10 shows one instance where ants are moving from nest to food which also indicates the release of pheromones during motion and also ants are moving forward in the path having pheromones.

If there are multi-path from the nest to food, then the amount of pheromones in the path determines which path to select by the ants to reach the food. If there are two paths from the nest to food, and the two paths have the equal length, then ants select both the path simultaneously. This is because in both the path amount of pheromones is same. So the probability of selecting any one path is also same.

Figure 11 shows two equal paths for the ants from nest to food. Since path length is equal so the amount of pheromones at the two paths are also equal. So the amount of ant in both the path is same. Figure 12 shows two unequal paths from nest to food. Since path-1 is shortest, so the amount of pheromones are more, and the probability of selecting the path-1 is more.

Nest

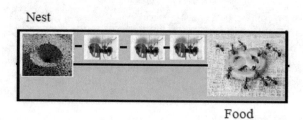

Food

Fig. 10 One instance of ACO

Path-1

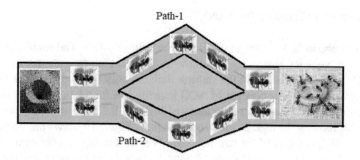

Path-2

Fig. 11 Two equal multipath for ACO

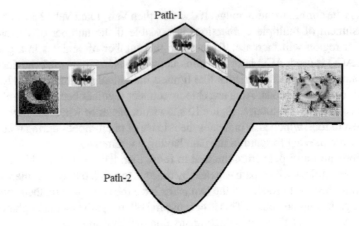

Fig. 12 Two unequal multipath for ACO

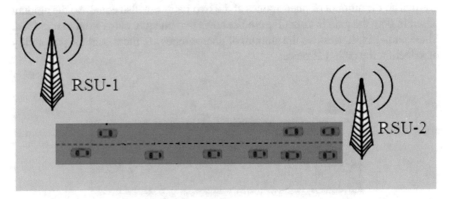

Fig. 13 Motion of vehicles in VANET

Modelling ACO Scenario for VANET

VANET scenario is also the same as the scenario of ACO. The mobile nodes in VANET i.e. vehicles always try to follow the shortest path from source to destination. Moving of the vehicle in the shortest path not only reduces the delay but also consumes less fuel. The concept of ACO is really helpful when there are multiple paths from source to destination. In terms of VANET, RSU keeps track of a number of vehicles moving through this path per unit time. Figure 13 shows one scenario of VANET which consists of the two-lane road. Vehicles in lane-1 communicate with RSU-1 to determine whether to move in lane-1 or not. Once the vehicle will move in lane-1, the information present in the RSU-1 is updated. A similar process is going on for the vehicles moving in lane-2.

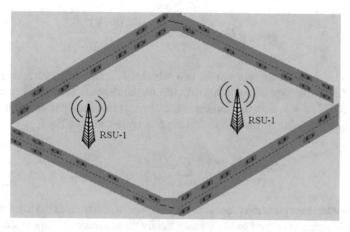

Fig. 14 Motion of vehicles in a multipath road in VANET

Thus, based on the information available in the RSU, vehicle decides whether to move in the path or alternate path. Figure 14 shows the two path scenario for the motion of vehicles.

Since, both the path have equal length, so the probability of moving the vehicles, in both the path will be the same. Another concept of ACO is the evaporation of pheromones released by the ant. Once the food has been obtained from the location, then the path is not used by the ant. This is possible because of the evaporation of pheromones. This helps ant not to move in that path later. Rana et al. proposed different ACO equations for VANET [51]. These equations are as follows.

$$LCS = \frac{\Delta t_{i,j}}{t_{max}} \qquad (4)$$

In Eq. 4, LCS is the link connectivity stability between the nodes i and j. $\Delta t_{i,j}$ is the time for which connectivity between node i and j will remain. This connectivity time is dependent on the speed of the node i,j and also depends on the direction of motion of the vehicles. t_{max} is the constant maximum time for which connectivity link between any two nodes will be present in the routing table. The calculation of pheromone deposit is represented by the Eq. 5

$$\Delta \phi_{i,j} = P_{Recvmsg} + LCS \qquad (5)$$

In the Eq. 5, $P_{Recvmsg}$ is the probability of receiving message between the node i and j. The value of $P_{Recvmsg}$ depends on the distance between the nodes i and j and also on the signal fading. If node i and j are within the communication range, then the value of $P_{Recvmsg}$ is more. $P_{Recvmsg}$ can be estimated by using Nakagami fading model [66]. This model is helpful for the dynamic behavior of VANET in urban and highway region [67]. Nakagami fading model for VANET is represented by Eq. 6.

$$P_{Recvmsg}(D, R) = e^{-m(\frac{D}{R})^2} \sum_{i=1}^{m} \frac{(m(\frac{D}{R})^2)^{i-1}}{(i-1)!} \qquad (6)$$

In Eq. 6, D is the distance between the two vehicles, R is the communication range. m is the fading parameter whose value depends on the network scenario. For example, if the value of m = 3, then this indicates that there is fast fading of the signal. Thus, substituting the value of Eqs. 4 and 6 in Eq. 5 is represented as Eq. 7.

$$\Delta\phi_{i,j} = e^{-m(\frac{D}{R})^2} \sum_{i=1}^{m} \frac{(m(\frac{D}{R})^2)^{i-1}}{(i-1)!} + \frac{\Delta t_{i,j}}{t_{max}} \qquad (7)$$

The pheromone evaporation process is very important, as it avoids the local optimum problem. Correia et al. proposed mathematically the pheromone evaporation process [68]. The pheromone evaporation mechanism is represented by Eq. 8.

$$\phi = (1 - \rho)\phi \qquad (8)$$

From Eq. 8, ρ is the evaporation rate for the pheromone. ϕ is the pheromone level associated with a path. The Eq. 8 shows that the pheromone level gradually decreases in a path. The path selection for a vehicle from the set of multi-paths is determined by Algorithm 1.

Algorithm: Path selection for a vehicle from multi-path.

Algorithm 1 Path determination for a vehicle from a junction

1: **Input**: Information like vehicle count, TTL for each path in the database;
2: **Output**: Determination of path for the movement of the vehicles;
3: **while** Position of vehicle is near the junction **do**
4:     Sends the request to RSU along with its speed.
5:     After receiving the request RSU compares the TTL value to select the path.
6:     **if** $TTL_{path1} > TTL_{path2}$ **then**
7:         Select path2
8:     **else if** $TTL_{path1} < TTL_{path2}$ **then**
9:         Select path1
10:     **else**
11:         Compare vehicle count
12:         **if** $Vehcount_{path1} > Vehcount_{path2}$ **then**
13:             Select path2
14:         **else if** $Vehcount_{path1} > Vehcount_{path2}$ **then**
15:             Select path1.
16:         **else**
17:             Select any path randomly.
18:         **end if**
19:     **end if**
20: **end while**
21: Update TTL and vehicle count for the requested vehicle in the selected path.
22: STOP

Complexity Analysis

The complexity of the proposed algorithm is dependent on the network scenario i.e. the number of junction (RSU) present in the network. If in the network n number of junctions are present, then the time complexity of the proposed algorithm is $\mathcal{O}(n)$.

5 Result

This section of the chapter discusses, the simulation set up and also the simulation result which includes the optimization technique ACO.

5.1 Simulation Set up

The proposed work is simulated using network simulator (NS2), mobility of the vehicles are implemented using SUMO, and the analysis of the result is done by the AWK scripts. Network scenario which consists of roads, junctions, number of vehicles, traffic lights, and path of the motion of the vehicles is described by SUMO. Part of the network scenario is shown in Fig. 15. The network scenario created by SUMO is integrated by NS-2, to generate the trace file and NAM file. NAM file is the network animation file to visualize the network. Using AWK script file, required data are extracted. The simulation set up is shown in Table 3. For the establishment of multiple routes by ACO, the information of the vehicles is stored in the open database called MongoDB. The operations with the database are done by node-red.

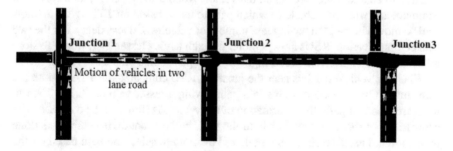

Fig. 15 Part of Network Scenario from SUMO

Table 3 Simulation network set up parameters

Parameters	Values
Simulation area	2000 m × 2000 m
Distance between junction	800 m
Communication range	200 m
Size of packet	512 bytes
Speed of the vehicles	30–50 km/hr
Number of vehicles	5–85
Number of Junction	9
Number of lanes	2
Propagation model	Two ray ground
Mac protocol	IEEE 802.11p
Queue	PriQueue
Queue length	50
Simulation time	100 s

5.2 Simulation Results

Initially, the determination of the path for a vehicle from the two paths is done. Establishment of multiple paths from source to destination by ACO is implemented in VANET. In VANET, determination of the multiple paths is done by RSUs. RSUs have a database which consists some information about the vehicle i.e. vcount, speed, TTL, path(either path 1 or path 2) for two-path connection from the source to destination. TTL refers to the sum of the time required by all the vehicles to move from one junction to another junction. When the vehicle reaches the junction, then RSU will communicate with the vehicle to which path to move based on TTL value. Always vehicle moves in the path having less number of vehicles and less delay. For the two path communication, RSUs always prefer the path having lesser TTL value. To store information about the vehicle, MongoDB—an open source database is used.

When the vehicle reaches near the junction, RSU compares the TTL of both the path. Inform the vehicle to move in a path having a lesser value of TTL. Then in the corresponding path, the information of the vehicle is stored in the database. The comparison of the TTL value of both the path in the MongoDB database is done by node-red. Figure 16 shows the node-red diagram to select the path based on the lesser value of TTL.

After the comparison of the TTL value, RSU transmits the message to move either in the path1 or path2. The output of node-red wiring is obtained in the serial monitor. Figure 17 shows the output of node-red wiring diagram.

Figure 18 shows the instance of mongoDB database.

Thus, pheromone in ACO is similar to the TTL value in the database. The only difference is, ants select the path having a greater pheromone, but vehicles select the

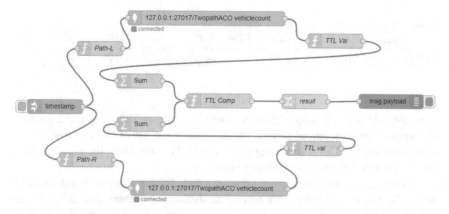

Fig. 16 Node-red wiring diagram for the selection of path

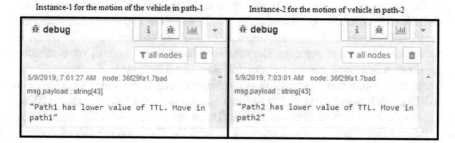

Fig. 17 Output of node-red

Fig. 18 Instance of the MongoDB

Table 4 End-to-end delay with respect to distance

Distance (m)	200	400	600	800	1000	1200	1400	1600
End-to-end delay (μs)	90	93	95	99	103	106	110	114

path having lesser TTL value in order to have a lesser delay. TTL value in the database also helps in the selection of the multi-path. Another concept of ACO i.e. evaporation of pheromone in ACO is similar to deletion of TTL value from the junction when the vehicle reaches near another junction.

After establishing multiple paths between source and destination using the concept ACO, the end to end delay with respect to distance is presented in Table 4. The end-to-end delay for the proposed work is compared with the two existing VANET routing protocol i.e. Geographic Source Routing (GSR) and Anchor Based Street Traffic Aware Routing (A-STAR). Figure 19 shows the comparison between end-to-end delay versus Distance of the proposed work with GSR and A-STAR keeping the number of vehicles in the network constant. In GSR, the transmission of information occurs with the help of vehicles. The message is transmitted through vehicles using the shortest distance between the vehicles. So with an increase in distance and with frequent reception and transmission of information, end-to-end delay increases. For A-STAR, end-to-end delay is initially less as compared to GSR. But after 1200m, end-to-end delay is more than that of GSR. This is because for A-STAR transmission of information depends upon the density of vehicles. Since the number of vehicles is constant, so with increases in distance, the density of vehicles decreases within a particular region. This increases end-to-end delay for A-STAR. But the proposed work tries to move the vehicles in the multipath with uniform density. Since both the path have uniform density so transmission of packet takes place in both the paths which occurs in parallel and reduces the end to end delay. Thus, the proposed work has less end-to-end delay.

Figure 20 shows the comparison between end-to-end delay vs the number of vehicles for the proposed work with the existing VANET protocol i.e. GSR and A-STAR keeping the distance between the source and destination constant. Figure 20 indicates that with an increase in the number of vehicles, the end-to-end delay decreases for GSR, A-STAR, and for the proposed work. This is because by keeping the distance between source and destination fixed and by increasing the number of vehicles the network gap decreases. The communication between the nodes takes place hop by hop without vehicle carrying delay. Since GSR depends on the shortest distance between the nodes to reach the destination, so the number of hops is more. Thus, due to frequent reception and transmission, end-to-end delay is more for GSR as compared to the other two protocols.

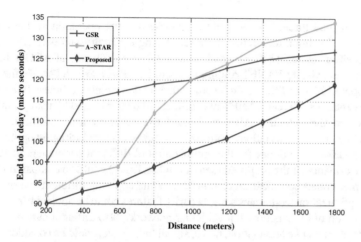

Fig. 19 End-to-End delay versus Distance

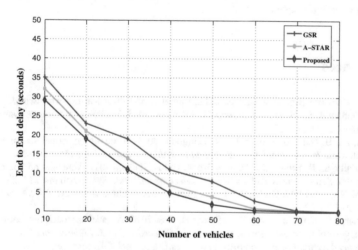

Fig. 20 End-to-End delay versus No. of vehicles

6 Conclusion and Future Scope

Ant Colony optimization and particle swarm optimization techniques are the efficient optimization techniques of swarm intelligence. In the proposed chapter, both ACO and PSO are considered in the vehicular network for the enhancement of the performance of VANET. Different emergency messages are classified and PSO is proposed for the transmission of emergency message like road accident information, information of the location where the fire is caught, etc. from the source location to destination location. Establishment of multiple routes from source to destination with uniform density of vehicles is proposed using ACO. The proposed idea of ACO is implemented through a MongoDB database and the operation on the database is performed using a tool called node-red. Using the multi-path network scenario, ACO is simulated using SUMO and NS2 and the performance of ACO is better as compared to existing VANET routing protocol i.e. GSR and A-STAR. In the future, the proposed idea of PSO could be simulated by determining a fitness function. An experimental test on real VANET platform could be performed by using transceiver modules in the vehicles. Also, we could use other swarm optimization techniques like artificial bee colony optimization, fuzzy bacterial foraging optimization, spyder monkey techniques for the optimization of different parameters in VANET. The performance of the proposed idea of PSO and ACO could be compared with the above-mentioned optimization techniques. While applying different swarm intelligence optimization techniques some security features could be applied for the secure transmission of information in a vehicular network. Also different network impairment parameters like attenuation, noise, signal fading, etc. could be considered in the routing protocol while optimizing the network parameters using swarm intelligence optimization techniques.

References

1. https://www.statista.com/statistics/281134/number-of-vehicles-in-use-worldwide/.
2. Abdelhamid, S., H. Hassanein, and G. Takahara. 2015. Vehicle as a resource (vaar). *IEEE Network* 29 (1): 12–17.
3. Das, H., B. Naik, B. Pati, and C.R. Panigrahi. 2014. A survey on virtual sensor networks framework. *International Journal of Grid Distribution Computing* 7 (5): 121–130.
4. Zeadally, S., R. Hunt, Y.S. Chen, A. Irwin, and A. Hassan. 2012. Vehicular ad hoc networks (vanets): Status, results, and challenges. *Telecommunication Systems* 50 (4): 217–241.
5. Bhoi, S.K., and P.M. Khilar. 2013. Vehicular communication: A survey. *IET Networks* 3 (3): 204–217.
6. Sarkar, J.L., C.R. Panigrahi, B. Pati, and H. Das. 2016. A novel approach for real-time data management in wireless sensor networks. In *Proceedings of 3rd international conference on advanced computing, networking and informatics*. pp. 599–607. Springer.
7. Li, F., and Y. Wang. 2007. Routing in vehicular ad hoc networks: A survey. *IEEE Vehicular technology magazine* 2 (2): 12–22.
8. Nagaraj, U., M. Kharat, and P. Dhamal. 2011. Study of various routing protocols in vanet. *IJCST* 2 (4): 45–52.
9. Paul, B., M. Ibrahim, M. Bikas, and A. Naser. 2012. Vanet routing protocols: Pros and cons. arXiv preprint arXiv:1204.1201.
10. Perkins, C.E., and P. Bhagwat. 1994. Highly dynamic destination-sequenced distance-vector routing (dsdv) for mobile computers. In *ACM SIGCOMM computer communication review*. vol. 24, pp. 234–244. ACM.
11. Clausen, T., and P. Jacquet. 2003. Optimized link state routing protocol (olsr). Tech. rep.
12. Johnson, D.B., and D.A. Maltz. 1996. Dynamic source routing in ad hoc wireless networks. In *Mobile computing*, pp. 153–181. Springer.
13. Perkins, C., E. Belding-Royer, and S. Das. 2003. Ad hoc on-demand distance vector (aodv) routing. Tech. rep.
14. Yu, F., Y. Li, F. Fang, and Q. Chen. 2007. A new tora-based energy aware routing protocol in mobile ad hoc networks. In *2007 3rd IEEE/IFIP international conference in central Asia on internet*. pp. 1–4. IEEE.

15. Khiavi, M.V., S. Jamali, and S.J. Gudakahriz. 2012. Performance comparison of aodv, dsdv, dsr and tora routing protocols in manets. *International Research Journal of Applied and Basic Sciences* 3 (7): 1429–1436.
16. Lochert, C., H. Hartenstein, J. Tian, H. Fussler, D. Hermann, and M. Mauve. 2003. A routing strategy for vehicular ad hoc networks in city environments. In *IEEE IV2003 intelligent vehicles symposium*. Proceedings (Cat. No. 03TH8683). pp. 156–161. IEEE.
17. Karp, B., and H.T. Kung. 2000. Gpsr: Greedy perimeter stateless routing for wireless networks. In *Proceedings of the 6th annual international conference on Mobile computing and networking*, pp. 243–254. ACM.
18. Naumov, V., and T.R. Gross. 2007. Connectivity-aware routing (car) in vehicular ad-hoc networks. In *IEEE INFOCOM 2007-26th IEEE international conference on computer communications*, pp. 1919–1927. IEEE.
19. Seet, B.C., G. Liu, B.S. Lee, C.H. Foh, K.J. Wong, and K.K. Lee. 2004. A-star: A mobile ad hoc routing strategy for metropolis vehicular communications. In *International conference on research in networking*. pp. 989–999. Springer.
20. Luo, Y., W. Zhang, and Y. Hu. 2010. A new cluster based routing protocol for vanet. In *2010 Second international conference on networks security, wireless communications and trusted computing*. vol. 1, pp. 176–180. IEEE.
21. Santos, R., R. Edwards, and A. Edwards. 2004. Cluster-based location routing algorithm for vehicle to vehicle communication. In *Proceedings 2004 IEEE radio and wireless conference (IEEE Cat. No. 04TH8746)*, pp. 39–42. IEEE (2004).
22. Xia, Y., C.K. Yeo, and B.S. Lee. 2009. Hierarchical cluster based routing for highly mobile heterogeneous manet. In *2009 international conference on network and service security*, pp. 1–6. IEEE.
23. Ibrahim, K., M.C. Weigle, and M. Abuelela. 2009. p-ivg: Probabilistic inter-vehicle geocast for dense vehicular networks. In *VTC Spring 2009-IEEE 69th vehicular technology conference*, pp. 1–5. IEEE.
24. Joshi, H.P., et al. 2007. *Distributed robust geocast: A multicast protocol for inter-vehicle communication*.
25. Rahbar, H., K. Naik, A. Nayak. 2010. Dtsg: Dynamic time-stable geocast routing in vehicular ad hoc networks. In *2010 The 9th IFIP annual mediterranean Ad Hoc networking workshop (Med-Hoc-Net)*, pp. 1–7. IEEE.
26. Durresi, M., A. Durresi, and L. Barolli. Emergency broadcast protocol for inter-vehicle communications. In *11th international conference on parallel and distributed systems (ICPADS'05)*. vol. 2, pp. 402–406. IEEE.
27. Korkmaz, G., E. Ekici, F. Özgüner, and Ü. Özgüner. 2004. Urban multi-hop broadcast protocol for inter-vehicle communication systems. In *Proceedings of the 1st ACM international workshop on Vehicular ad hoc networks*, pp. 76–85. ACM.
28. Nagaraj, U., and P. Dhamal. 2012. Broadcasting routing protocols in vanet. *Network and Complex Systems* 1 (2): 13–19.
29. Kumar, V., S. Mishra, and N. Chand. 2013. Applications of vanets: Present & future. *Communications and Network* 5 (01): 12.
30. Barba, C.T., M.A. Mateos, P.R. Soto, A.M. Mezher, and M.A. Igartua. 2012. Smart city for vanets using warning messages, traffic statistics and intelligent traffic lights. In *2012 IEEE intelligent vehicles symposium*, pp. 902–907. IEEE (2012).
31. Khekare, G.S., and A.V. Sakhare. 2013. A smart city framework for intelligent traffic system using vanet. In *2013 international mutli-conference on automation, computing, communication, control and compressed sensing (iMac4s)*. pp. 302–305. IEEE.
32. Chen, R., W.L. Jin, and A. Regan. 2010. Broadcasting safety information in vehicular networks: Issues and approaches. *IEEE Network* 24 (1): 20–25.
33. Bhoi, S.K., D. Puthal, P.M. Khilar, J.J. Rodrigues, S.K. Panda, and L.T. Yang. 2018. Adaptive routing protocol for urban vehicular networks to support sellers and buyers on wheels. *Computer Networks* 142: 168–178.

34. Zhao, J., and G. Cao. Vehicle-assisted data delivery in vehicular ad hoc networks. *IEEE Transactions on Vehicular Technology*. v57 i3 1922.
35. Tufail, A., M. Fraser, A. Hammad, K.K. Hyung, and S.W. Yoo. 2008. An empirical study to analyze the feasibility of wifi for vanets. In *2008 12th international conference on computer supported cooperative work in design*, pp. 553–558. IEEE.
36. Popoola, S.I., O.A. Popoola, A.I. Oluwaranti, A.A. Atayero, J.A. Badejo, and S. Misra. 2017. A cloud-based intelligent toll collection system for smart cities. In *International conference on next generation computing technologies*, pp. 653–663. Springer.
37. Chaurasia, B.K., and S. Verma. 2014. Secure pay while on move toll collection using vanet. *Computer Standards & interfaces* 36 (2): 403–411.
38. Senapati, B.R., P.M. Khilar, and N.K. Sabat. 2019. An automated toll gate system using vanet.
39. Milojevic, M., and V. Rakocevic. 2014. Distributed road traffic congestion quantification using cooperative vanets. In *2014 13th annual Mediterranean Ad Hoc networking workshop (MED-HOC-NET)*, pp. 203–210. IEEE.
40. Younes, M.B., and A. Boukerche. 2013. Efficient traffic congestion detection protocol for next generation vanets. In *2013 IEEE international conference on communications (ICC)*, pp. 3764–3768. IEEE.
41. Panayappan, R., J.M. Trivedi, A. Studer, and A. Perrig. 2007. Vanet-based approach for parking space availability. In *Proceedings of the fourth ACM international workshop on Vehicular ad hoc networks*, pp. 75–76. ACM.
42. Senapati, B.R., R.R. Swain, and P.M. Khilar. 2020. Environmental monitoring under uncertainty using smart vehicular ad hoc network. In *Smart intelligent computing and applications*, pp. 229–238. Springer.
43. Safi, Q.G.K., S. Luo, L. Pan, W. Liu, and G. Yan. 2018. Secure authentication framework for cloud-based toll payment message dissemination over ubiquitous vanets. *Pervasive and Mobile Computing* 48: 43–58.
44. Kennedy, J.: Swarm intelligence. In *Handbook of nature-inspired and innovative computing*, pp. 187–219. Springer.
45. Bonabeau, E., D.d.R.D.F. Marco, M. Dorigo, G. Théraulaz, and G. Theraulaz, et al. 1999. Swarm intelligence: From natural to artificial systems. No. 1, Oxford University Press.
46. Nayak, J., B. Naik, A. Jena, R.K. Barik, and H. Das. 2018. Nature inspired optimizations in cloud computing: Applications and challenges. In *Cloud computing for optimization: Foundations, applications, and challenges*, pp. 1–26. Springer.
47. Kennedy, J. 2010. Particle swarm optimization. *Encyclopedia of Machine Learning*, pp. 760–766.
48. Dorigo, M., and M. Birattari. 2010. *Ant colony optimization*. Springer.
49. Karaboga, D., and B. Basturk. 2007. A powerful and efficient algorithm for numerical function optimization: Artificial bee colony (abc) algorithm. *Journal of Global Optimization* 39 (3): 459–471.
50. Passino, K.M. 2010. Bacterial foraging optimization. *International Journal of Swarm Intelligence Research (IJSIR)* 1 (1): 1–16.
51. Rana, H., P. Thulasiraman, R.K. Thulasiram. 2013. Mazacornet: Mobility aware zone based ant colony optimization routing for vanet. In *2013 IEEE congress on evolutionary computation*, pp. 2948–2955. IEEE.
52. Sahoo, R.R., R. Panda, D.K. Behera, and M.K. Naskar. 2012. A trust based clustering with ant colony routing in vanet. In *2012 third international conference on computing, communication and networking technologies (ICCCNT'12)*, pp. 1–8. IEEE.
53. Shoaib, M., and W.C. Song. 2012. Data aggregation for vehicular ad-hoc network using particle swarm optimization. In *2012 14th Asia-Pacific network operations and management symposium (APNOMS)*, pp. 1–6. IEEE.
54. Toutouh, J., and E. Alba. 2012. Parallel swarm intelligence for vanets optimization. In *2012 Seventh international conference on P2P, parallel, grid, cloud and internet computing*, pp. 285–290. IEEE.

55. Zukarnain, Z.A., N.M. Al-Kharasani, S.K. Subramaniam, and Z.M. Hanapi. 2014. Optimal configuration for urban vanets routing using particle swarm optimization. In *International conference on artificial intelligence and computer science*, pp. 1–6.
56. Kaiwartya, O., and S. Kumar. 2014. Geocasting in vehicular adhoc networks using particle swarm optimization. In *Proceedings of the international conference on information systems and design of communication*, pp. 62–66. ACM.
57. Das, H., A.K. Jena, J. Nayak, B. Naik, and H. Behera. 2015. A novel pso based back propagation learning-mlp (pso-bp-mlp) for classification. In *Computational intelligence in data mining*, Vol. 2, pp. 461–471. Springer.
58. Fekair, M.E.A., A. Lakas, and A. Korichi. 2016. Cbqos-vanet: Cluster-based artificial bee colony algorithm for qos routing protocol in vanet. In *2016 International conference on selected topics in mobile & wireless networking (MoWNeT)*, pp. 1–8. IEEE.
59. Zhang, X., X. Zhang, and C. Gu. 2017. A micro-artificial bee colony based multicast routing in vehicular ad hoc networks. *Ad Hoc Networks* 58: 213–221.
60. Zhang, X., and X. Zhang. 2017. A binary artificial bee colony algorithm for constructing spanning trees in vehicular ad hoc networks. *Ad Hoc Networks* 58: 198–204.
61. Baskaran, R., M.S. Basha, J. Amudhavel, K.P. Kumar, D.A. Kumar, and V. Vijayakumar. A bio-inspired artificial bee colony approach for dynamic independent connectivity patterns in vanet. In *2015 International conference on circuits, power and computing technologies [ICCPCT-2015]*, pp. 1–6. IEEE.
62. Panigrahi, C.R., J.L. Sarkar, B. Pati, and H. Das. 2015. S2s: A novel approach for source to sink node communication in wireless sensor networks. In *International conference on mining intelligence and knowledge exploration*, pp. 406–414. Springer.
63. Hafeez, K.A., L. Zhao, B. Ma, and J.W. Mark. 2013. Performance analysis and enhancement of the dsrc for vanet's safety applications. *IEEE Transactions on Vehicular Technology* 62 (7): 3069–3083.
64. Kenney, J.B. 2011. Dedicated short-range communications (dsrc) standards in the united states. *Proceedings of the IEEE* 99 (7): 1162–1182.
65. Morgan, Y.L. 2010. Notes on dsrc & wave standards suite: Its architecture, design, and characteristics. *IEEE Communications Surveys & Tutorials* 12 (4): 504–518.
66. Killat, M., and H. Hartenstein. 2009. An empirical model for probability of packet reception in vehicular ad hoc networks. *EURASIP Journal on Wireless Communications and Networking* 2009: 4.
67. Rubio, L., J. Reig, and N. Cardona. 2007. Evaluation of nakagami fading behaviour based on measurements in urban scenarios. *AEU-International Journal of Electronics and Communications* 61 (2): 135–138.
68. Correia, S.L.O., J. Celestino, and O. Cherkaoui. Mobility-aware ant colony optimization routing for vehicular ad hoc networks. In *2011 IEEE wireless communications and networking conference*, pp. 1125–1130. IEEE.

Development of Fast and Reliable Nature-Inspired Computing for Supervised Learning in High-Dimensional Data

Hiram Ponce, Guillermo González-Mora, Elizabeth Morales-Olvera and Paulo Souza

Abstract Machine learning and data mining tasks in big data involve different nature of inputs that typically exhibit high dimensionality, e.g. more than 1,000 features, far from current acceptable scales computing in one machine. In many different domains, data have highly nonlinear representations that nature-inspired models can easily capture, outperforming simple models. But, the usage of these approaches in high-dimensional data are computationally costly. Recently, artificial hydrocarbon networks (AHN)—a supervised learning method inspired on organic chemical structures and mechanisms—have shown improvements in predictive power and interpretability in contrast with other well-known machine learning models, such as neural networks and random forests. However, AHN are very time-consuming that are not able to deal with big data until now. In this chapter, we present a fast and reliable nature-inspired training method for AHN, so they can handle high-dimensional data. This training method comprises a population-based meta-heuristic optimization with defined both individual encoding and objective function related to the AHN-model, and it is also implemented in parallel-computing. After benchmark performing of population-based optimization methods, grey wolf optimization (GWO) was selected. Our results demonstrate that the proposed hybrid GWO-based training method for AHN runs more than $1400x$ faster in high-dimensional data, without loss of predictability, yielding a fast and reliable nature-inspired machine learning model. We also present a use case in assisted living monitoring, i.e. human fall classification comprising 1,269 features from sensor signals and video recordings, with

H. Ponce (✉) · G. González-Mora · E. Morales-Olvera
Facultad de Ingeniería, Universidad Panamericana, Augusto Rodin 498,
03920 Ciudad de México, México
e-mail: hponce@up.edu.mx

G. González-Mora
e-mail: 0147901@up.edu.mx

E. Morales-Olvera
e-mail: 0177650@up.edu.mx

P. Souza
Faculty UNA of Betim, Av. Gov. Valadares, 640 Centro, 32510-010 Betim, MG, Brazil
e-mail: goldenpaul@informatica.esp.ufmg.br

© Springer Nature Switzerland AG 2020
M. Rout et al. (eds.), *Nature Inspired Computing for Data Science*,
Studies in Computational Intelligence 871,
https://doi.org/10.1007/978-3-030-33820-6_5

this proposed training algorithm to show its implementation and performance. We anticipate our new training algorithm to be useful in many applications like medical engineering, robotics, finance, aerospace, and others, in which big data is essential.

1 Introduction

Machine learning is continuously releasing its power in a wide range of applications, big data allows these algorithms to make more accurate and timely predictions; but it carries a cost that involves challenges for machine learning in model scalability and distributed computing. Machine learning and data mining tasks in big data includes different nature of inputs that typically exhibit high dimensionality, e.g. more than 1,000 features, far from current acceptable scales computing in one machine [5].

In general, the definition of high dimensional data is linked to a direct relationship between a large number of samples being evaluated with a large number of dimensions. Other authors define this high dimensionality when the number of features exceeds the number of samples of the problem [12]. In this chapter, we consider high dimensional data when the ratio between the number of samples and the number of dimensions is large enough, and also that the number of dimensions exceeds 1,000 as commented in [5].

Machine learning can be characterized by the nature of learning feedback, target of learning tasks, and timing of data availability. In many different domains, data have highly nonlinear representations that nature-inspired models can easily capture, outperforming simple models. Particularly, examples of nature-inspired computing and machine learning models dealing with complex problems in different dimensional data can be found in works like: [11] in which authors identified diabetes using J48 and Naive Bayes models and also provided tools for exploring and understanding the output rules; the authors in [51] identified intruders in networks using an improvement of C4.5 decision trees over a dataset comprising 39 features and 21,606 samples; a PSO-based model in a multi-layer perceptron network was studied in [10] for classification purposes over datasets ranging from 4 to 13 features and less than 350 samples; authors in [32] addressed large-scale cloud computing problems based on nature-inspired computing methods. In most of the cases, the nature-inspired computing methods are used for classification in medium dimensional data, and a few works use nature-inspired optimization methods to improve the data-driven models.

Meta-heuristics optimization algorithms are useful in a wide range of applications because of their characteristics [27, 60]: (i) based on simple ideas for easier implementation; (ii) capable to find optimal neighborhood solution; and (iii) they can be implemented in different areas.

Population-based methods, in which possible solutions are represented by individuals, are typically inspired on social behavior of animals and other entities. These meta-heuristic optimization algorithms generates a set of candidates that traverse the search space to find optimal solutions. At each step, these individuals share information to follow the best solution found so far. Mechanisms like cooperation, organization and decentralization of the individuals allow emerging intelligence in the population, so near-optimal solutions can be reached out [13, 31, 57].

Some recent bio-inspired population-based optimization algorithms are: particle swarm optimization (PSO), grey wolf optimization (GWO), and bat (BAT) algorithm, among others. PSO is a heuristic global optimization method and algorithm based on swarm intelligence. It aims to simulate the biological behavior of fish schooling and bird flocking [56]. The BAT algorithm was based on the echolocation features of microbats [59]. BAT increase the variety of the solutions in the population using a frequency tuning technique, simultaneously, during the search process, it uses automatic zooming to balance exploration and exploitation by imitating differences in pulse emission rates and bats loudness when searching for prey. Other method, GWO, mimics the social leadership and hunting technique and of grey wolves. The main steps of the GWO algorithm are based on: (i) track, chase and approach the prey; (ii) pursue, encircle and harass; (iii) attack [31].

Recently, artificial hydrocarbon networks (AHN)—a supervised learning method inspired on organic chemical structures and mechanisms—have shown improvements in predictive power and interpretability in contrast with other well-known machine learning models, such as neural networks and random forests [37, 45]. However, AHN are very time-consuming and are not able to deal with big data until now. Big data is mainly characterized by the amount of information that can be process, the speed of data generation and the variety in data involved. Existing machine learning algorithms need to be adapted to profit the advantages of big data and process more information efficiently.

The AHN model uses a gradient based learning algorithm that, due to its complexity, hinders the scalability of the model. Regarding scalability, as the input dimensionality of real world applications increases, the execution time of the training phase grows dramatically. This is due to the hierarchical training that finds optimal parameters in molecules and then updates the center of molecules. In any case, this training method coupled with the AHN model has shown favorable results and properties in regression and classification problems; such as stability, robustness, packaging data and parameter interpretability [45].

Specifically, optimization algorithms have been incorporated to machine learning models for their ability to approximate high dimensional functions with measurable accuracy and precision with a gradient free optimization approach; making it easier to compute on real world problems. From these algorithms, meta-heuristic approaches have yielded especially accurate results for classification and regression tasks [30].

Meta-heuristic algorithms tend to increase computational cost if they are based on population exploration instead of trajectory movement of a single solution. To minimize the impact of meta-heuristic approximations on the training phase, distributed computing frameworks and parallelism strategies have been implemented. While distribution strategies look to divide the tasks in numerous executors that compute the algorithm on a portion of the data set; parallelism strategies look to utilize all available cores to enable multi-threaded execution [5].

In this chapter, we present a fast and reliable nature-inspired training method for AHN, so they can handle high-dimensional data. This training method comprises a population-based meta-heuristic optimization with defined both individual encoding and objective function related to the AHN-model, and it is also implemented in parallel-computing. After benchmark performing of population-based optimization methods, GWO was selected. Our results demonstrate that the proposed training method for AHN runs more than $1400x$ faster in high-dimensional data, without loss of predictability, yielding a fast and reliable nature-inspired machine learning model. We also present a use case in human fall classification with this proposed training algorithm to show its implementation and performance. The data set employed contains $28,356$ samples with $1,269$ features, considered high dimensional data [5, 12] (e.g. ratio of 22.34 between samples and features).

The contribution of this work relies on the hybrid training method for AHN using GWO that outperforms the training execution time $1400x$ faster than the original training algorithm, and how this hybrid algorithm is implemented in two stages of learning, i.e. training centers of molecules and molecular parameters independently (as described in Sect. 3). The noticeable decrease in execution time allows the proposed model to operate in several areas where data flow is high, and answers need to be taken in real-time. We anticipate our new training algorithm to be useful in many applications like medical engineering, robotics, finance, aerospace, and others, in which big data is essential. Furthermore, it is important to highlight that this is the first time that the training algorithm for AHN is improved for fast and reliable learning data-driven models. Also, this is the first time that AHN can handle high dimensional data (c.f. Table 7) due to the implementation of a parallel-based meta-heuristic optimization approach, in contrast to the original training algorithm that cannot deliver an output response after 24 h of execution, as shown later in the case study.

The rest of the chapter is organized as follows. Section 2 presents an overview of artificial hydrocarbon networks, from the inspiration on nature to the implementation on machine learning. Then, Sect. 3 presents our proposal for fast and reliable training method for artificial hydrocarbon networks based on the grey wolf optimization. Section 4 shows the performance of the proposed method and Sect. 5 describes a case study on human fall classification using artificial hydrocarbon networks in a large data set. After that, some trends and applications of artificial hydrocarbon networks are presented in Sect. 6. Finally, we conclude the chapter.

2 Artificial Hydrocarbon Networks

Based on artificial organic networks, Ponce and Ponce [41] proposed a supervised machine learning technique known as Artificial Hydrocarbon Networks (AHN). The main purpose of this method is to model data using the inspiration of carbon networks, replicating the chemical rules related to organic molecules which can describe the structure and behavior of data [38, 45].

2.1 Inspiration from Organic Compounds

In chemistry, the simplest unit is the atom, matter is made up of a set of them conditioned by the environment so that interactions between atoms are possible after optimum configurations with minimal energy. The interaction of atoms entails to chemical bonds that join them together to form molecules and mixtures of them. Therefore, the structure of compounds takes into account not only the set of atoms but also the ways they are attracted to each other. Energy minimization and geometric configuration are important in the structural organization. Nevertheless, different chemical behavior results from the relationship between different atoms. In that sense, stability, organization and multi-functionality are the base of organic compounds inspiration to develop a machine learning. Following, these nature observations are briefly described.

2.1.1 Stability

Stability is the property of compounds that allows them to maintain their geometric configurations. Organic compounds, in particular, are the most stable in nature. This stability depends on electronegativity, which is a chemical property of atoms [8]. It describes the tendency of electron attraction among atoms. In regard to electronegativity values, the basic atoms related to organic compounds have the highest ones.

2.1.2 Organization

In addition, organic compounds also maintain minimal inner energy, based on the principle of the ground-state [24]. Both organic and inorganic compounds tend to minimize the energy between atoms, getting the ground-state. Organic compounds follow a chemical heuristic rule that claims three levels of energy in the formation of configurations to preserve this energy minimization [45]. The first level considers the ground-state principle for molecules and atoms bonds through electron sharing. The next level considers chemical reactions to combine unstable molecules and form more complex compounds. Finally, the last level describe chemical balance interaction in mixtures derived by the relation of two or more compounds. Through this, in compounds (and in complexity) the three energy levels intuitively design a hierarchical organization.

2.1.3 Multi-functionality

The functional group of organic compounds describes their chemical behavior. Each group consist on organic molecules made of a carbon atom and other atoms from a set of compounds, such as: hydrogen, oxygen, nitrogen, phosphorus, fluorine, sulfur, chlorine, iodine, bromine and astatine [24]. For instance, the organic compounds most studied associated with these functional groups are [6]: hydrocarbons, alcohols, aldehydes, amines, ketones, carboxylic acids, carbohydrates, polymers, lipids, proteins, amino acids and nucleic acids. It is evident from these studies that changes in the atoms related inside compounds may give different behaviors. In addition, the various combinations of eleven atoms may lead into more than 20 million known organic compounds [24]. This ability to build a wide variety of compounds from very few atoms results in multi-functionality and diversity.

2.2 Overview of Artificial Hydrocarbon Networks Method

From the foregoing, the supervised learning method of AHN loosely simulates how the hydrocarbon molecules interact. For readability, Table 1 summarizes the description of the chemical-based terms used in the AHNs technique described below and their meanings [40, 45].

The AHN method consists only of hydrogen and carbon elements which can be combined with up to one and four atoms, respectively (see Fig. 1). Simple molecules, also known as the basic unit information are called CH-molecules, and they are formed by linkage of one carbon atom with $1 \leq k \leq 4$ hydrogen atoms, denoted as CH_k. One molecule generally models a chunk of data in its parameters (hydrogen and carbon atoms) and arrangement. Figure 1 shows the configuration of a molecule as its structural representation, which leads to their chemical behavior. Mathematically, the behavior φ of a molecule with k hydrogen atoms is expressed as in (1); where,

Table 1 Description of the chemical terms used in AHN and their computational meanings

Chemical terminology	Symbols	Meaning
Environment	$x = (x_1, \ldots, x_n)$	(features) data inputs in $\mathbb{R}^n$
Ideal behavior	y	(target) data outputs, solution of mixtures
Atoms	H_i, σ	(parameters) basic structural units or properties
Molecules	$\varphi(x)$	(functions) basic units of information
Compounds	$\psi(x)$	(composite functions) complex units of information made of m molecules
Mixtures	$S(x)$	(linear combinations) combination of compounds
Stoichiometric coefficient	α_t	(weights) definite ratios in mixtures
Molecular centers	μ_j	(vector) representation of the centered data in molecules
Energy	E_0, E_j	(loss function) value of the error between target and estimate values

Fig. 1 Structure of an artificial hydrocarbon network using saturated and linear chains of molecules. Inputs enter to the AHN model (mixture) made of different compounds, and it outputs the response of the model. A sample of a linear compound is depicted in the top of the image, showing the formation of m molecules. Each molecule is centered in μ_j that is a point in the input space

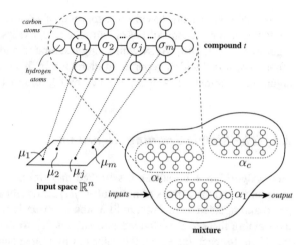

$\sigma \in \mathbb{R}^n$ is called the carbon value, $H_i \in \mathbb{R}^n$ is the i-th hydrogen atom attached to the carbon atom, and $x = (x_1, \ldots, x_n)$ is the input vector with n features.

$$\varphi(x, k) = \sum_{r=1}^{n} \sigma_r \sum_{i=0}^{k \leq 4} H_{ir} x_r^i \tag{1}$$

Two or more unsaturated molecules, i.e. those with hydrogen atoms less than 4, can be joined together. In AHN, this new structure is called compound. Different compounds have been defined in literature [44–46]. The simplest one is the saturated and linear chain of m molecules. It is composed structurally of two CH_3 molecules and $(m - 2)$ CH_2 molecules, as shown in Fig. 1. The behavior ψ of a saturated-and-linear compound is defined as (2); where, φ_j is the behavior of the jth associated molecule that represents a subset Σ_j of the input x such that $\Sigma_j = \{x \mid \arg\min_j (x - \mu_j) = j\}$, and $\mu_j \in \mathbb{R}^n$ is the center of the jth molecule [37, 40]. In fact, $\Sigma_{j_1} \cap \Sigma_{j_2} = \emptyset$ if $j_1 \neq j_2$.

$$\psi(x) = \begin{cases} \varphi_1(x, 3) & x \in \Sigma_1 \\ \varphi_2(x, 2) & x \in \Sigma_2 \\ \cdots & \cdots \\ \varphi_{m-1}(x, 2) & x \in \Sigma_{m-1} \\ \varphi_m(x, 3) & x \in \Sigma_m \end{cases} \tag{2}$$

Mixtures $S(x)$ are formed by the interaction of compounds with each other in certain ratios α_t, so-called weights or stoichiometric coefficients. They are represented as set in (3); where, c shows the number of compounds in the mixture and α_t, the weighted factor of the t-th compound [45].

$$S(x) = \sum_{t=1}^{c} \alpha_t \psi_t(x) \tag{3}$$

An AHN is broadly based on a mixture of compounds (see Fig. 1) each one computed using a chemical-based heuristic rule, expressed in the so-called AHN-algorithm [41, 45, 46]. AHN with a single saturated-and-linear compound [37, 38, 40, 41, 44, 45] is widely reported in literature. In this chapter, the simplest training algorithm (original-AHN) is represented by Algorithm 1.

2.3   The Simple Training Algorithm

First, Algorithm 1 initializes a saturated and linear compound structure with m molecules, and the centers of molecules $\{\mu_j\}$ randomly set. The compound is computed and updated as follows, until a stop criterion is reached. For each molecule, the training dataset Σ is divided into subsets Σ_j so that every input x is close to μ_j. Then, for each molecule, the values of hydrogen and carbon are independently computed, using the QR-factorization for solving the least squares estimates (LSE) method, and the associated error E_j between the output response of the jth molecule and the actual targets y of the jth subset is calculated. In the case a subset is empty, it will require a new position, i.e. update their centers. This relocation is done by simply randomly changing the center of the empty subset close to one molecule with

---

Algorithm 1 (original-AHN) Training algorithm for a single saturated-and-linear compound of AHN.

Input: the training data set $\Sigma = (x, y)$, the number of molecules in the compound $m \geq 2$ and the learning rate $0 < \eta < 1$.
Output: the trained compound ψ.

1: Create the structure of a new compound with m molecules.
2: Randomly initialize a set of molecular centers, μ_j.
3: **while** not reach a stop criterion **do**
4:     **for** $j = 1$ to m **do**
5:         $\Sigma_j \leftarrow$ do a subset of Σ using μ_j.
6:         $\{H_j, \sigma_j\} \leftarrow$ calculate molecular parameters of φ_j using LSE method.
7:         $E_j \leftarrow$ compute the error in molecule.
8:     **end for**
9:     **for** $j = 1$ to m **do**
10:         **if** $\Sigma_j = \emptyset$ **then**
11:             $\mu_j \leftarrow$ relocate the center of molecule near to other μ_k such that E_k is very large.
12:         **end if**
13:     **end for**
14:     Update all centers using $\mu_j = \mu_j - \eta(E_{j-1} - E_j)$.
15: **end while**
16: Update the behavior of compound ψ using all φ_j already calculated.
17: **return** ψ.

---

a big error. After the relocation, the molecules center are updated using a gradient descent approach with learning rate $0 < \eta < 1$, previously selected, as seen in (4) with $E_0 = 0$. The last step, involves an update of the compound ψ with the behaviors of molecules calculated so earlier. More detailed description of the original-AHN can be found in [45].

$$\mu_j \leftarrow \mu_j - \eta(E_{j-1} - E_j) \tag{4}$$

This method is based on a general framework, i.e. artificial organic networks (AON) [45], which aims to create artificial organic models, including: (i) a graph structure analogous to their physical properties, (ii) a mathematical model behavior associated to their chemical properties, and (iii) a training algorithm that finds appropriate the appropriate model parameters. Packaging information in modules (molecules) is one of the main characteristics of AON [45]. The packages are organized and optimized using chemical energy-based heuristic mechanisms as defined in the training algorithm. Both artificial organic networks and the algorithms designed under this framework, e.g. AHNs, enable: modularity and organization of data, inheritance of packaging information, and structural stability of data packages. The properties of AHN method are useful when considering regression and classification problems, as shown in following:

1. *Stability*—the AHN-algorithm ensures minimal changes in the output response when there is a slightly change in the inputs [45], encouraging the usage of the AHNs as a supervised learning method.
2. *Robustness*—it considers that the algorithm is able to deal with uncertain data and noise, AHN acts as a filtering system. Literature reports examples, such as: audio filtering [44], as classifiers in human activity recognition systems [38, 40], and ensembles of AHNs with fuzzy logic as intelligent control systems [42, 43, 47].
3. *Packaging data*—it alouds to compute molecular structures into the algorithm, similar data with alike capabilities is clustered together. The data is not only packed by its features, but also by its tendency [45].
4. *Parameter interpretability*—it means that AHN can be useful to extract features or partially understand underlying information applying different coefficients and weights. As an example, the AHN-algorithm has been used in approaches related with facial recognition when using its parameters as metadata information [45], also it has been implemented in a cancer diagnosis system to describe its decision-making tasks [37].

More detailed information related with properties and comparisons among other supervised learning methods can be found in [37, 40, 45, 46].

2.4 Issues on the Training Algorithm

The first-AHN training algorithm has reported good performance in predictive power for low-dimensional input spaces, but high training time in the model for computing appropriate parameters. These issues are mainly related to: (i) the hierarchical training, which requires first finding optimal parameters in molecules and then updating the center of molecules; and (ii) data split, performed whenever the molecules center is recalculated. In that sense, as suggested below, other mechanisms for AHN training are required.

3 Parallel Grey Wolf Training Algorithm for Artificial Hydrocarbon Networks

In this section, we present the proposed parallel artificial hydrocarbon networks (P-AHN), an algorithm for fast and robust training of AHN. This proposal is a hybrid method that comprises two key components: (i) parallel-computing based meta-heuristic optimization for near optimal convergence of centers of molecules, and (ii) molecular parameter (i.e. carbon and hydrogen values) computation using the LSE method. Figure 2 shows the block diagram of the proposed training algorithm.

The main procedure of the proposed P-AHN training algorithm is the computation of the centers of molecules using a meta-heuristic optimization. For that, we propose to benchmark three population-based methods for possible inclusion: particle swarm optimization (PSO), grey wolf optimization (GWO) and bat optimization (BAT). Later on, we consider parallel-computing since these are population-based methods. Following, details about these optimization methods and the proposed training algorithm are presented.

3.1 Population-Based Meta-Heuristic Optimization Methods

Exact optimization algorithms are excellent in solving problems where mathematical modeling is not as complicated or when the amount of data is not high. However, there is incredibly costly when optimization methods are applied to large volumes of data, hampering the search for local minima and doing the exhaustive search for optimal solutions impractical. To solve these problems, there are meta-heuristic optimization algorithms that are capable of solving complex problems with large data volume or with a large number of candidate solutions. They act for global surveys of complex problems with global exploration and local exploitation capacity, allowing the steps of the algorithm to be simpler and more precised [4].

For Glover and Kochenberger [15], these methods of optimization stand out by the incorporation of methods and devices that avoid unnecessary confinements in the

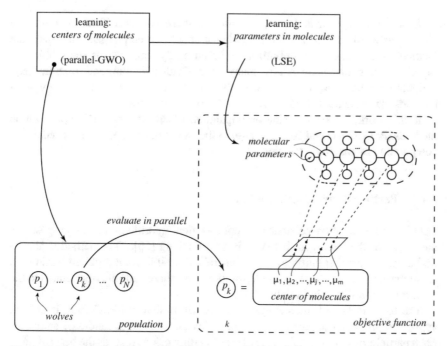

Fig. 2 Main components of the parallel artificial hydrocarbon networks training method. Two key components interact as follows. (i) GWO method calculates the best suitable center of molecules, each time an individual is evaluated, this runs on parallel, and (ii) parameters in molecules are optimized using LSE, all inside the objective function. At the end, the wolf α is the best solution found

search space, especially when the amount of data is very high. These methods work using specific knowledge of the problem, as heuristics controlled by high level strategies. Current techniques incorporate advanced approaches in the research experience to assist in the task of optimizing the problem.

Meta-heuristic optimization techniques can act differently in problem-solving. In general, they are classified as follows [4]:

- *Nature-inspired against non-nature inspired*—it gave rise to those transforming observations from nature into computational strategies, such as: particle swarm optimization (PSO) [23], grey wolf optimization (GWO) [31], bat optimization [58], genetic algorithms (GA) [16], ant colony optimization (ACO) [13], among others. These approaches act in a way inspired on groups of living beings to perform the main optimization tasks.
- *Population-based against single point search*—these are groups of algorithms classified accordingly to the search mechanisms for optimal solutions. Single point search works with one solution at any time. Population-based methods are those based on swarms or a set of individuals, searching multiple candidate solutions in parallel-fashion.

Beheshti and Shamsuddin [4] explain that these two classifications cover the algorithms most used in the literature, but it should be emphasized that there are methods of meta-heuristic optimization: based on dynamic versus static objective function (e.g., guided local search), many versus single neighborhood structures (e.g. variable neighborhood search), and finally the memory usage versus memory-less methods that execute their tasks based on Markov processes.

In this work, we are focusing on population-based meta-heuristic optimization methods (PSO, GWO and BAT) because of the nature of parallelism inherited on the computation of candidate solutions.

3.1.1 Particle Swarm Optimization

PSO is a population-based stochastic optimization algorithm that was proposed as the model of the intelligent behavior in bird flocking [56]. This method is able to find a near-optimal solution to an unconstrained optimization problem based on a set of particles that encodes possible solutions and shares information related to the performance of that value solutions so far.

The basic PSO with momentum, also known as standard-PSO [60], is defined as follows. It sets a population of N particles that represent candidate solutions. Each particle $p_i = (p_1, \ldots, p_j, \ldots, p_D)$ of D dimensions records the best solution, p_{best}, found at iteration t. In addition, the whole algorithm records the best particle solution, g_{best}, found at iteration t.

To move particles, each one has a velocity v_i. In the standard-PSO, at each iteration t, the velocity of each particle is updated using the rule of (5), where w refers to the inertia weight, and $\alpha_1 > 0$ and $\alpha_2 > 0$ are the cognitive and social coefficients.

Velocity values are bounded to velocity limits. Then, the position of all particles are updated using (6). Once again, the resultant positions are bounded by position limits.

$$v_{i,j}(t+1) = w \times v_{i,j}(t) + \alpha_1 \times r^1_{i,j} \times (p_{best,j}(t) - p_{i,j}(t)) +$$
$$+ \alpha_2 \times r^2_{i,j} \times (g_{best,j}(t) - p_{i,j}(t)) \tag{5}$$

$$p_{i,j}(t+1) = p_{i,j}(t) + v_{i,j}(t+1) \tag{6}$$

3.1.2 Grey-Wolf Optimization

In 2014 Mirjalili et al. [31] proposed GWO that is a meta-heuristic optimization based on the behavioral nature of the wolf pack, mimicking the hierarchy of wolf hunting. The levels of the relevancy of wolves determine the form and attacks on food hunting, and the algorithm was inspired precisely in this approach. The optimization approach can simulate computationally the main steps performed by the gray wolves during the search for food: searching for prey, encircling prey, and attacking prey.

The modeling of the solution is inspired by the organization of the pack of gray wolves, where the main highlight is the alpha solution, which represents the most appropriate solution to the problem. In the pack, the alpha wolf (α) is responsible for the significant decisions about attack order and organization when feeding a prey. The next candidate solutions follow the same logic, the second most effective being represented by beta (β) and the third most efficient is called the delta (δ). Finally, the other candidate solutions are called omegas (ω). The structure in the search for the best optimization results follows this hierarchy, being the alpha solution that dictates the rhythm of the execution of the activities [31].

The encircling prey stage is defined as (7) and (8) [31], where $\vec{A}$ and $\vec{C}$ are coefficient vectors, $\vec{X}_p$ is the position vector of the prey, $\vec{X}$ indicates the position vector of a grey wolf and t indicates the current iteration.

$$\vec{D} = |\vec{C} \cdot \vec{X}_p(t) - \vec{X}(t)| \tag{7}$$

$$\vec{X}(t+1) = \vec{X}_p(t) - \vec{A} \cdot \vec{D} \tag{8}$$

The coefficient vectors are calculated for the problem in (9), where $\vec{a}$ has elements that decay from 2 to 0 and r_1 and r_2 are random vectors between [0, 1] [31].

$$\vec{A} = 2\vec{a} \cdot \vec{r}_1 - \vec{a} \tag{9}$$

For the hunting stage, the reference is always the alpha wolf, aided not mandatory by the beta wolf. As well as complex problems that do not know the optimal value of the solution, the prey can be in places not known. To solve this problem, a candidate function representing the alpha wolf is used to unravel the solution, thus allowing three best solutions to the problem to be obtained. With these resolutions in hand, it is possible to update the other wolves with the best references obtained, thus simulating the natural behavior of the pack in search of prey. This step is modeled as (10), (11) and (12) [31].

$$\vec{D}_\alpha = |\vec{C}_1 \cdot \vec{X}_\alpha - \vec{X}|, \vec{D}_\beta = |\vec{C}_2 \cdot \vec{X}_\beta - \vec{X}|, \vec{D}_\delta = |\vec{C}_3 \cdot \vec{X}_\delta - \vec{X}| \tag{10}$$

$$\vec{X}_1 = \vec{X}_\alpha - \vec{A}_1 \cdot (\vec{D}_\alpha), \vec{X}_2 = \vec{X}_\beta - \vec{A}_2 \cdot (\vec{D}_\beta), \vec{X}_3 = \vec{X}_\delta - \vec{A}_3 \cdot (\vec{D}_\delta) \tag{11}$$

$$\vec{X}(t+1) = \frac{\vec{X}_1 + \vec{X}_2 + \vec{X}_3}{3} \tag{12}$$

The stage of nature hunting ends when the prey is captured and stops moving. With the algorithm it works the same way. When the alpha approaches the optimal solution, all the other elements also approach to confirm the optimal solution of the problem. The search for new prey also follows this behavior [31].

The GWO technique, although recent, has already been employed in the resolution of several complex problems, such as those that were solved by Emari et al. [14] that uses the optimization technique to aid in the selection of features. Precup et al. used the technique to reduce the parametric sensitivity of fuzzy control systems [48].

3.1.3 Bat Optimization

BAT algorithm was proposed by Yang and Gandomi in [58] aiming to solve engineering optimization problems using a meta-heuristic optimization algorithm inspired by nature. The idea was to propose a new meta-heuristic method that explores the echo-based location behavior that bats have, and preliminary studies have shown that it is quite promising.

This algorithm determines the positions x and velocities v of bats moved in terms of the frequencies f emitted by them. The updating rules are expressed as (13), (14) and (15), where $F\beta$ is a uniform random number between [0, 1] and x_* is the current global best solution, obtained after comparing all the solutions among all the n bats.

$$f_i = f_{\min} + (f_{\max} - f_{\min})F\beta \tag{13}$$

$$v_i^t = v_i^{t-1} + \left(x_i^{t-1} - x_*\right) f_i \tag{14}$$

$$x_i^t = x_i^{t-1} + v_i^t \tag{15}$$

In general, the positions of the bats are updated in the initial stages of the algorithm. If a random number generated is greater than the pulse emission rate, a new solution is generated around the best current global solution. Thus, the algorithm can be seen as in (16) [58], where $\varepsilon \in [-1, 1]$ is a random number and $A^t = \langle A_i^t \rangle$ is the average loudness of all the bats in iteration t. The loudness A_i and rate of pulse emission r_i are updated as the iterations proceed, normally using a geometric rate A_{rate} and r_{rate}, respectively.

$$x_{\text{new}} = x_{\text{old}} + \varepsilon A^t \tag{16}$$

This natural approach is also very much applied in solving complex optimization problems. Recent work by Jia et al. [22] reported the usage of BAT on oil pollution image segmentation in addition with neural networks, the work of Zhang et al. [59] that uses this technique to aid in short-term load forecasting. The work of Talbi et al. [53] proposed a fuzzy controller rule-base optimized through BAT. Lastly, Anter and Ali [1] used BAT and other optimization methods to improve the performance of models working with databases from medical diagnostics.

3.2 Meta-Heuristic Optimization for Artificial Hydrocarbon Networks, and Benchmark

For our P-AHN proposal, we decided to experiment which of the three meta-heuristic optimization methods is the best one to compute the centers of molecules in AHN. Thus, we firstly introduce the general procedure to hybridize a population-based meta-heuristic optimization in the training AHN, and then we compared the experimental results.

3.2.1 General Hybrid Meta-Heuristic Optimization and AHN Training Algorithm

We set two components to hybridize meta-heuristic optimization methods into the AHN training algorithm: (i) the individual encoding and (ii) the objective function.

For particle encoding, we are simply representing the set of centers of molecules μ_j as the dimensions of the individuals p_i, such that $p_i = (\mu_1, \ldots, \mu_m)$ for all $i = 1, \ldots, N$ (see Fig. 2). On the other hand, the objective function $\mathtt{obj}$ is proposed such that hydrogen and carbon values ($w = \{\sigma_r, H_{ir}\}$) are calculated using LSE, while fixing the center of molecules μ_j. After that, the error E between the response of the molecular behavior and the targets is computed. This error value results to be the value of the objective function f_i, such that $f_i = \mathtt{obj}(p_i, \Sigma, m)$. Algorithm 2 shows the objective function $\mathtt{obj}$ proposed.

Algorithm 2 Objective function $\mathtt{obj}$ for individual evaluation.

Input: the individual $p_i = (\mu_1, \ldots, \mu_m)$, the training data set $\Sigma = (x, y)$ and the number of molecules in the compound $m \geq 2$.
Output: the overall error E in the model and the trained compound ψ.

1: Create the structure of a new compound with m molecules.
2: Set the centers of molecules using p_i.
3: **for** $j = 1$ to m **do**
4:     $\Sigma_j \leftarrow$ do a subset of Σ using μ_j.
5:     $w \leftarrow$ calculate the weights of φ_j using LSE method.
6:     $\{H_j, \sigma_j\} \leftarrow$ compute the molecular parameters using $w_{ir} = \sigma_r H_{ir}$ and $H_{1r} = 1$.
7:     $E_j \leftarrow$ compute the error in molecule.
8: **end for**
9: Compute the overall error $E = \sum_j E_j$.
10: Build the behavior of compound ψ using all φ_j already calculated.
11: **return** E, ψ.

3.2.2 Comparative Analysis of P-AHN Using Different Meta-Heuristic Optimization Methods

We conducted a comparative analysis on the performance of the hybrid meta-heuristic optimization methods (PSO, GWO and BAT) including in the AHN training algorithm. Two experiments were set up for this purpose.

The first experiment is a regression problem that comprises a synthetic data set with one feature x and one target y. The output target follows the expression shown in (17), for $x \in [-1, 1]$ and $\epsilon \sim \mathcal{N}(0, 0.05)$.

$$y = \begin{cases} \arctan \pi x + \epsilon & x < 0.1 \\ \sin \pi x + \epsilon & 0.1 \leq x < 0.6 \\ \cos \pi x + \epsilon & x \geq 0.6 \end{cases} \tag{17}$$

Table 2 Setup of parameters in the meta-heuristic optimization methods

Method	Parameters
PSO	Population $= 25$
	Max. iterations $= 10,000$
	$\alpha_1 = 2.0$
	$\alpha_2 = 2.0$
	$w = 0.7$
GWO	Population $= 25$
	Max. iterations $= 10,000$
BAT	Population $= 25$
	Max. iterations $= 10,000$
	$A_{rate} = 0.9$
	$r_{rate} = 0.9$
	$f_{min}, f_{max} = [0, 100]$

The second experiment is also a regression problem of a synthetic data set with two features, x_1 and x_2, and one target y. In this case, targets are formulated as defined in (18).

$$y : \begin{cases} x_1 = \cos t \\ x_2 = t \\ t \in [0, 15] \end{cases} \tag{18}$$

In both experiments, we changed the number of samples to $q = 1000$, $10,000$ and $1,00,000$, generating the data points uniformly. Also, in each of the experiments, we varied the number of molecules ($m = 3$, 10 and 20) to be considered in the AHN training algorithm.

For these experiments, we implemented PSO, GWO and BAT separately. We also configured the individual encoding and the objective function as shown in the previous section. Table 2 shows the meta-parameters fixed in the meta-heuristic optimization methods. For training purposes, we split the data into 70% training and 30% testing. Then, we ran a 10-fold cross-validation for training the different models. We measured both the execution time in training and the root-mean squared error (RMSE). Tables 3 and 4 summarize the performance of the hybrid methods and the original AHN training algorithm as baseline (with maximum 10000 iterations).

As observed the execution time, in training, from Tables 3 and 4, we found that the hybrid method with PSO performed well with one feature and a small set of samples, but it did not maintain the ranking while the number of molecules, features and samples increase. Moreover, the hybrid method with BAT only performed well with one feature and a large set of samples, regardless to the number of molecules. In addition, the hybrid method with GWO performed better in most of the cases, preferably with two features and with any number of molecules and samples. It is remarkable to say that most of the execution times using any of the hybrid methods outperforms, $51.9x$ in average, to the original training algorithm (i.e. Algorithm 1) when the number of features is two, $q = 100000$ and $m = 20$. Also, RMSE in the

Table 3 Performance on one-feature synthetic data using serialized computation. It reports mean (standard deviation) of the 10-fold cross validation training process. Bold text represents the best model evaluation

Algorithm	Molecules	1000 samples		10000 samples		100000 samples	
		Execution time (s)	RMSE	Execution time (s)	RMSE	Execution time (s)	RMSE
Original	3	3.585(0.152)	0.318(0.021)	78.879(2.623)	0.182(0.127)	874.175(0.112)	0.279(0.007)
PSO		**2.450(0.014)**	0.064(0.037)	**25.042(0.204)**	0.057(0.042)	244.774(280.593)	0.088(0.044)
GWO		2.518(0.076)	**0.057(0.043)**	25.243(0.431)	**0.057(0.027)**	**241.925(14.418)**	**0.060(0.0151)**
BAT		2.967(0.120)	0.106(0.054)	27.337(0.566)	0.115(0.051)	257.568(9.932)	0.086(0.043)
Original	10	24.530(0.208)	0.188(0.093)	212.511(0.185)	0.143(0.001)	12911.508(0.631)	0.145(0.001)
PSO		7.855(0.104)	**0.019(0.011)**	76.554(0.677)	**0.029(0.008)**	785.175(292.971)	0.037(0.017)
GWO		**7.794(0.022)**	**0.019(0.011)**	**76.397(0.704)**	0.032(0.014)	774.692(14.418)	**0.030(0.015)**
BAT		8.790(0.060)	**0.019(0.011)**	81.181(3.438)	0.296(0.021)	**627.944(2.752)**	0.233(0.122)
Original	20	153.001(0.228)	0.228(0.010)	2003.975(95.533)	0.285(0.002)	42636.388(3.479)	0.309(0.031)
PSO		**15.279(0.047)**	0.014(0.001)	158.172(17.552)	**0.022(0.005)**	1678.895(275.983)	**0.025(0.009)**
GWO		15.392(0.056)	**0.013(0.001)**	**149.722(0.448)**	0.031(0.008)	1534.918(9.869)	**0.025(0.009)**
BAT		17.880(0.314)	**0.013(0.001)**	160.318(4.056)	0.034(0.045)	**1236.760(0.441)**	0.320(0.179)

Table 4 Performance on two-feature synthetic data using serialized computation. It reports mean (standard deviation) of the 10-fold cross validation training process. Bold text represents the best model evaluation

Algorithm	Molecules	1000 samples		10000 samples		100000 samples	
		Execution time (s)	RMSE	Execution time (s)	RMSE	Execution time (s)	RMSE
Original	3	6.575(0.184)	0.021(0.013)	132.390(102.969)	0.672(0.012)	1454.700(7.779)	0.810(0.002)
PSO		2.295(0.126)	**0.001(0.001)**	21.612(0.240)	**0.666(0.011)**	202.940(4.079)	**0.706(0.001)**
GWO		**2.288(0.226)**	**0.001(0.001)**	**21.234(0.148)**	0.682(0.011)	**202.741(3.866)**	**0.706(0.001)**
BAT		2.289(0.147)	0.004(0.001)	21.236(0.185)	0.688(0.007)	202.848(3.192)	**0.706(0.001)**
Original	10	42.560(0.2937)	0.005(0.001)	333.693(2.312)	2.408(0.002)	16937.100(18282.500)	6.020(0.006)
PSO		6.984(0.339)	**0.001(0.001)**	63.217(0.483)	2.258(0.406)	616.400(7.405)	6.117(0.278)
GWO		**6.869(0.467)**	**0.001(0.001)**	**62.108(0.080)**	2.302(0.576)	**615.460(1.473)**	5.817(0.488)
BAT		7.162(0.210)	0.004(0.002)	67.825(0.545)	**2.210(0.365)**	663.966(1.275)	**2.490(0.259)**
Original	20	245.092(147.018)	0.006(0.001)	2448.400(1868.300)	2.601(0.001)	64899.300(75450.100)	12.003(0.003)
PSO		14.571(2.163)	**0.001(0.001)**	123.519(0.166)	**2.233(0.196)**	1221.111(9.107)	11.778(0.302)
GWO		**12.999(0.094)**	**0.001(0.001)**	**122.178(1.091)**	2.379(0.298)	**1214.729(2.470)**	11.584(0.470)
BAT		14.073(0.179)	**0.001(0.001)**	132.140(0.253)	2.290(0.398)	1320.090(8.650)	12.499(0.225)

hybrid methods improved the performance in contrast to the original AHN training algorithm. Based on the above analysis, we decided to hybrid GWO with the AHN training algorithm.

3.3 Parallel Training Method

We consider a parallel computing on the population-based meta-heuristic optimization method to speed up the performance of the whole P-AHN training algorithm.

It works as follows. When the meta-heuristic optimization method (i.e. GWO in this proposal) runs, the inner calls of the objective function run in parallel (e.g. Algorithm 2, lines 3–8). Moreover, the objective function evaluations of individuals are also proposed to run in parallel (e.g. Algorithm 3, lines 4–7 and lines 11–14). This parallel computing is inspired in the work of [34] in which authors presented a parallel-PSO contribution for initialization, objective function evaluation, finding the best solution (local and global), and updating of position and velocity of particles. A similar parallel-PSO was described in [26]. To this end, our proposal accelerates the computation highly, as shown later in the experiments. Algorithm 3 depicts the implementation of the whole P-AHN including the proposed GWO.

Algorithm 3 P-AHN with GWO.

Input: the training data set $\Sigma = (x, y)$, the number of molecules in the compound $m \geq 2$ and the batch size $\beta > 0$.
Output: the trained compound ψ.

1: Set the lower and upper bounds of wolves' position, lb and ub, respectively.
2: Set the number of wolves N in the population, and parameter settings of GWO.
3: Initialize positions of wolves p_i.
4: Initialize a, A and C.
5: **parfor** $i = 1$ to N **do**
6:     $f_i = \mathrm{obj}(p_i, \Sigma_i, m)$.
7: **end parfor**
8: Compute the best wolves X_α, X_β and X_δ using f_i.
9: **while** stop criteria not reached **do**
10:     Update positions of wolves p_i using (12).
11:     Update a, A and C.
12:     **parfor** $i = 1$ to N **do**
13:         $f_i = \mathrm{obj}(p_i, \Sigma_i, m)$.
14:     **end parfor**
15:     Update the best wolves X_α, X_β and X_δ using f_i.
16: **end while**
17: Build the best AHN-model using $[f_{best}, \psi] = \mathrm{obj}(X_\alpha, \Sigma, m)$.
18: **return** ψ.

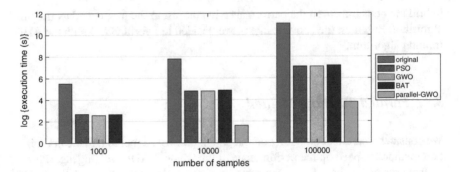

Fig. 3 Comparison among the training methods used in this work, for the worst case scenario implemented: two features, $m = 20$ and $q = 100000$. It shows the execution time values, in y-log scale, per training method

4 Experimental Results

We conducted the same experiments, as shown in Sect. 3.2.2, in order to analyze the performance of the proposed parallel P-AHN with GWO. We computed the execution time in training and the RMSE of the building models. Tables 5 and 6 summarize the evaluation metrics of the P-AHN with GWO against the original AHN training algorithm.

As shown in Tables 5 and 6, the P-AHN with GWO outperforms, in execution time, to the original AHN training algorithm. In terms of RMSE, the original AHN training algorithm performed better only when the number of molecules $m = 3$ and the number of samples $q = 10000$ and $q = 100000$, in contrast to the parallel P-AHN with GWO. But, the remaining cases, P-AHN with GWO outperforms in RMSE too. In this regard, the parallel P-AHN with GWO is $693.9x$ faster, in average, than the original training algorithm. Moreover, P-AHN with GWO is $1473.9x$ faster than the original AHN training algorithm in the worst scenario of $m = 20$ and $q = 100000$.

Figure 3 shows a visual comparison among the serialized hybrid methods employed, i.e. PSO, GWO and BAT, the parallel P-AHN with GWO, and the original AHN training algorithm; all in terms of the two-feature data set with $m = 20$ and $q = 100000$ (worst case). It shows the execution time in y-log scale in order to clearly compare the large values of the original AHN training algorithm. By visual inspection, it is easy to easy how the parallel P-AHN with GWO outperforms the other methods. In the same fashion, Fig. 4 depicts a visual comparison about the RMSE values. As observed, RMSE in all hybrid methods are similar to each other and better than the original AHN training algorithm, in low-dimensional data. In high-dimensional data, the performance of RMSE remains close in all the methods, except for the hybrid method with BAT algorithm. Lastly, the parallel P-AHN with GWO is slightly better in all the cases for different number of samples.

Table 5 Performance on one-feature synthetic data using parallel computation. It reports mean (standard deviation) of the 10-fold cross validation training process. Bold text represents the best model evaluation

Algorithm	Molecules	1000 samples		10000 samples		100000 samples	
		Execution time (s)	RMSE	Execution time (s)	RMSE	Execution time (s)	RMSE
Original	3	3.585(0.152)	0.318(0.021)	78.879(2.623)	0.182(0.127)	874.175(0.112)	0.279(0.007)
Parallel-GWO		**0.756(0.020)**	**0.080(0.058)**	**2.821(0.270)**	**0.146(0.054)**	**22.170(0.176)**	**0.127(0.040)**
Original	10	24.530(0.208)	0.188(0.093)	212.511(0.185)	**0.143(0.001)**	12911.508(0.631)	**0.145(0.001)**
Parallel-GWO		**1.084(0.075)**	**0.136(0.006)**	**3.549(0.058)**	0.214(0.048)	**30.783(0.111)**	0.198(0.036)
Original	20	153.001(0.228)	**0.228(0.010)**	2003.975(95.533)	**0.285(0.002)**	42636.388(3.479)	**0.309(0.031)**
Parallel-GWO		**1.178(0.097)**	0.253(0.017)	**4.986(0.072)**	0.357(0.035)	**43.630(0.472)**	0.369(0.057)

Table 6 Performance on two-feature synthetic data using parallel computation. It reports mean (standard deviation) of the 10-fold cross validation training process. Bold text represents the best model evaluation. These results show that the proposed training algorithm is faster and more reliable than the original training algorithm

Algorithm	Molecules	1000 samples		10000 samples		100000 samples	
		Execution time (s)	RMSE	Execution time (s)	RMSE	Execution time (s)	RMSE
Original	3	6.575(0.184)	0.021(0.013)	132.390(102.969)	0.672(0.012)	1454.700(7.779)	0.810(0.002)
Parallel-GWO		**0.800(0.054)**	**0.004(0.001)**	**2.636(0.017)**	1.094(0.272)	**23.580(0.444)**	1.562(0.516)
Original	10	42.560(0.2937)	0.005(0.001)	333.693(2.312)	2.408(0.002)	16937.100(18282.500)	6.020(0.006)
Parallel-GWO		**0.914(0.037)**	**0.001(0.001)**	**3.639(0.117)**	**1.957(0.552)**	**31.357(0.247)**	**5.592(0.508)**
Original	20	245.092(147.018)	0.006(0.001)	2448.400(1868.300)	2.601(0.001)	64899.300(75450.100)	12.003(0.003)
Parallel-GWO		**1.039(0.030)**	**0.001(0.001)**	**5.080(0.050)**	1.991(0.565)	**44.030(0.226)**	**10.796(0.884)**

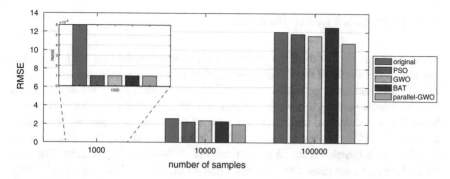

Fig. 4 Comparison among the training methods used in this work, for the worst case scenario implemented: two features, $m = 20$ and $q = 100000$. It shows RMSE values per training method

All the experiments (and the case study presented in the next section) were run in an Intel Core i7-7820HK at 4.3 GHz with two 8 GB in RAM (dual channel); and an NVIDIA GeForce GTX 1080–8192 MB, 1732 MHz in core with 2700 MHz in memory (GDDR5X) ForceWare 378.66.

5   Case Study on Human Fall Detection

To validate that our proposed P-AHN with GWO works in big data, we introduce this case study on the classification of multiple types of human falls and daily activities. For readability, fall detection monitoring is briefly presented. Then, the fall data set employed for this case study is described, and lastly experimentation and results are presented and discussed.

5.1   Fall Detection Classification

Falls are considered as an important health problem worldwide. Fall detection systems can alert when a fall occurs reducing the time in which the person obtains medical attention [61]. This issue is especially relevant in the elderly, since they are the most likely to trip, fall and sustain the most serious injuries [20]. The latest developments on this issue focus on fall recognition based on motion sensor data taken from wearable sensors or smartphones; typically using integrated gyroscopes and accelerometers [55]. Applications of classification methods such as support vector machines, random forests and k-nearest neighbors use data that has been previously filtered and divided into time windows to make pattern recognition easier by smoothing the motion data, thus enhancing performance of the classifier [17, 20, 61].

5.2   Description of the Data Set

For this case study, we used the public UP-Fall Detection data set [28]. It is multimodal in the sense of the different sources of information gathered to represent human falls and daily activities. It comprises of two data sets[1]: the consolidated and the feature sets. In this work, we consider the feature data set for experimentation that was collected and processed from 17 test subjects, measured from three sources: wearable sensors, ambient sensors and vision devices. The subjects were 9 males and 8 females from 18 to 24 years old without any impairment. The data set includes 11 activities: five different types of human falls and six simple daily activities: *falling forward using hands* (1), *falling forward using knees* (2), *falling forward backwards* (3), *falling sideward* (4) and *falling sitting on a missing chair* (5), *walking* (6), *standing* (7), *sitting* (8), *picking up an object* (9), *jumping* (10) and *laying* (11).

Data was collected from 14 sensors: six infrared sensors, two cameras, one electroencephalograph helmet and five wearable sensors located at the ankle, pocket, waist, neck and wrist of the subjects. The sampling rate of the final data set is 18 Hz [28]. Features extracted from the wearable and infrared sensors comprise 12 temporal features (mean, standard deviation, root mean square, maximal amplitude, minimal amplitude, median, zero-crossing, skewness, kurtosis, first quartile, third quartile and auto-correlation) and six frequency features (mean, median, entropy, energy, principal frequency and spectral centroid) [2, 3, 7, 9, 35, 49, 50], and 400 motion features from vision devices were computed as described below. A windowing process was done at intervals of 1 s with overlapping of 0.5 s. At the end, the final data set employed in this work comprises of 28,356 samples with 1,269 features (ratio of 22.34 between samples and features), and 11 label classes were considered.

In addition, the data set contains features that represent the relative motion of pixels between two consecutive images, obtained from the Horn and Schunck optical flow method, as reported in [28]. Using this information, in this work, we processed the features as follows: for each camera, we retrieved all image features inside a window. These features are the horizontal and vertical relative movements in the scenes, known as u and v respectively. These u and v components are two numeric matrices with the same size of the original images. For interpretability, we combined these two components resulting in the magnitude of the relative movement as shown in (19), where d is the resultant matrix of size equals to the original image.

$$d_{i,j} = \sqrt{u_{i,j}^2 + v_{i,j}^2} \qquad (19)$$

We, then, re-sized the resultant matrix d from 640×480 to 20×20 size. After that, we reshaped matrix d in a row vector of 400 elements. Lastly, all these row vectors from image features inside a window were averaged. Thus, a 400-row vector was obtained for each window, representing the features for images.

---

[1] Publicly available in http://sites.google.com/up.edu.mx/har-up/.

5.3 Human Fall Classification Using Artificial Hydrocarbon Networks

For this case study, we built an AHN-model using the proposed P-AHN training algorithm with GWO (Algorithm 3). We set $m = 5$ molecules, obtained from 10-fold cross-validation over the number of molecules from $m = \{3, \ldots, 10\}$ and measuring the classification error rate.

It is remarkable to say that AHN is mainly for regression problems. In this regard, we included a soft-max function [18] at the output of the AHN, so that classification can be performed easily. This soft-max function is expressed as (20), where the output $\hat{y} \in \mathbb{R}^o$ is a zero-vector of indexes with value 1 in the position that represents the class of the human activity/fall, and $\psi_i(x)$ is the i-th response of the AHN-model.

$$\hat{y}_i = \frac{\exp(\psi_i(x))}{\sum_{i=1}^{o} \exp(\psi_o(x))} \tag{20}$$

For final evaluation, we adopted the 10-fold cross-validation training approach mainly used in fall detection systems [29, 33]. At each fold, 70% of training and 30% of testing data were randomly selected from the data set. In addition, we built four well-known machine learning models recognized in fall detection systems [7, 21, 25, 29, 33, 54]: random forest (RF), support vector machines (SVM), multi-layer perceptron (MLP) and k-nearest neighbors (kNN). Five metrics were also evaluated [28]: *accuracy, precision, sensitivity, specificity* and F_1-*score*. Table 7 shows the performance of the machine learning models for fall detection. These results were sorted based on the F_1-score, a typical metric employed in fall detection, that measures the balance between precision and sensitivity [28].

It can be observed that AHN-model computes the best metrics in most of the cases, including F_1-score. MLP is the second best machine learning model, and it is interesting that SVM ranked the lowest. In terms of *accuracy* (97.02%), *sensitivity* (86.77%) and *specificity* (99.72%), the proposed training algorithm outperforms the

Table 7 Performance of the machine learning models for fall classification. Values are reported as mean (standard deviation). Bold numbers represent the best scores. The original AHN training algorithm did not reach any response after waiting 24 h

Method	F_1-score (%)	Accuracy (%)	Precision (%)	Sensitivity (%)	Specificity (%)
AHN (original)	–	–	–	–	–
AHN (parallel-GWO)	**76.68 ± 0.236**	**97.02 ± 0.005**	72.12 ± 0.067	**86.77 ± 0.121**	**99.72 ± 0.0001**
MLP	70.44(1.25)	94.32(0.31)	**76.78(1.59)**	67.29(1.42)	99.42(0.03)
RF	69.36(1.35)	95.09(0.23)	75.53(2.31)	66.23(1.11)	99.50(0.02)
kNN	60.51(0.85)	92.06(0.24)	68.82(1.61)	58.49(1.14)	99.19(0.02)
SVM	55.82(0.77)	91.16(0.25)	66.79(2.79)	53.82(0.70)	99.07(0.02)

H. Ponce et al.

results. Lastly, it is important to notice that the original AHN training algorithm did not reach any results after waiting 24 h in training execution time. In that sense, the proposed P-AHN with GWO highly improves the training performance of AHN.

This case study validated that our proposed P-AHN with GWO training algorithm can build AHN-models with a very competitive predictive power (e.g. in terms of *accuracy, precision, sensitivity, specificity* and F_1-*score*); but also, it can handle large amount of data (28356 sample-pairs of 1269 features and 11 output-vector classes).

6 Trends and Applications of Artificial Hydrocarbon Networks

Until now, AHN have been successfully applied to different problems and different learning tasks. However, there still are challenges and issues about AHN that have to be faced. Trends in the development and application of AHN can be listed as follows:

- New training algorithms for AHN are required for better computational performance mainly in time. For example: distributed-computing, parallel-computing, or different LSE methods.
- The next big step in AHN is parallel computing. In this regard, this proposal opens the possibility for processing big data analysis.
- Since the predictive power of AHN is well accurated, it is important to study other functions as kernels in molecules aiming to perform other approximations. In addition, other relationships among molecules within compounds might model complex nonlinearities in data.
- Hybrid approaches with AHN might improve solutions to very complex problems, such as: robotics, bussiness intelligence, healthcare, and others.
- Few efforts in dynamic modeling using AHN are reported in literature [36]. It is important to focus some research in this direction.
- Transfer learning using pre-defined molecules can be done, but more studies are necessary.
- Interpretability of machine learning models and especifically of AHN models, are of great importance as shown in the proposed molecule-based clustering for knowledge extraction. Automatic procedures for this task are required.

In terms of applications, literature reports many different implementations of AHN. Those can be classified as follows: function approximation and modeling [45]; robust human activity recognition systems [38, 40]; signal processing in denoising audio and face recognition [44, 45]; online advertising [39]; intelligent control systems for robotics [36, 43, 45, 47] and mechatronics [19, 45, 52]; bio/medical applications [37, 38, 40]; and, theoretical approaches such as hybrid fuzzy-molecular inference systems [43], and interpretability of the model [37].

7 Conclusions

In this chapter, we presented a hybrid AHN training algorithm with a population-based meta-heuristic optimization method, i.e. GWO, referred as P-AHN with GWO, to accelerate the training execution procedure in AHN, but also to maintain reliability on the predictive power of this supervised learning method. The methodology for this proposal consisted on analyzing and determining a suitable optimization method that improves the training performance in AHN. To do so, we split the training procedure to optimize the centers of molecules with the meta-heuristic algorithm while the molecular parameters are optimized in parallel with the LSE method.

Experimental results validated that the proposed P-AHN with GWO method increases the speed of AHN training 1400 times in contrast with the original training method. In addition, this method was implemented on the human fall classification case study to determine its feasibility in big data. After the experiments, we concluded that AHN can be built using the proposed method, against the original training algorithm that could not obtain a response in at least 24 h.

For future work, we are considering to analyze other meta-heuristic algorithms to hybridize the AHN training algorithm. In addition, other optimization techniques should be tested. Thus, we also anticipate that our new training algorithm will be useful in many applications like medical engineering, robotics, finance, aerospace, computer vision, and so many others.

References

1. Anter, A.M., and M. Ali. 2019. Feature selection strategy based on hybrid crow search optimization algorithm integrated with chaos theory and fuzzy c-means algorithm for medical diagnosis problems. *Soft Computing* 1–20.
2. Atallah, L., B. Lo, R. King, and G.Z. Yang. 2010. Sensor placement for activity detection using wearable accelerometers. In *2010 International conference on body sensor networks*, 24–29. IEEE.
3. Avci, A., S. Bosch, M. Marin-Perianu, R. Marin-Perianu, and P. Havinga. 2010. Activity recognition using inertial sensing for healthcare, wellbeing and sports applications: A survey. In *23rd International conference on architecture of computing systems (ARCS)*, 1–10. Hannover: Germany.
4. Beheshti, Z., and S.M.H. Shamsuddin. 2013. A review of population-based meta-heuristic algorithms. *International Journal of Advances in Soft Computing and its Applications* 5 (1): 1–35.
5. Bekkerman, R. 2012. *Scaling up machine learning*. Cambridge University Press.
6. Brown, W., C. Foote, B. Iverson, and E. Anslyn. 2011. *Organic chemistry*. Cengage Learning.
7. Bulling, A., U. Blanke, and B. Schiele. 2014. A tutorial on human activity recognition using body-worn inertial sensors. *ACM Computing Surveys (CSUR)* 46 (3): 1–33.
8. Carey, F., and R. Sundberg. 2007. *Advanced organic chemistry: Part A: Structure and mechanisms*. Springer.
9. Dargie, W. 2009. Analysis of time and frequency domain features of accelerometer measurements. In *2009 Proceedings of 18th International Conference on Computer Communications and Networks*, ICCCN 2009, 1–6. IEEE.

10. Das, H., A.K. Jena, J. Nayak, B. Naik, and H. Behera. 2015. A novel PSO based back propagation learning-MLP (PSO-BP-MLP) for classification. In *Computational intelligence in data mining-volume 2*, 461–471. Springer.

11. Das, H., B. Naik, and H. Behera. 2018. Classification of diabetes mellitus disease (DMD): Ad data mining (DM) approach. In *Progress in computing, analytics and networking*, 539–549. Springer.

12. Donoho, D.L., et al. 2000. High-dimensional data analysis: The curses and blessings of dimensionality. *AMS Math Challenges Lecture* 1 (2000): 32.

13. Dorigo, M., and M. Birattari. 2010. *Ant colony optimization*. Springer.

14. Emary, E., H.M. Zawbaa, and A.E. Hassanien. 2016. Binary grey wolf optimization approaches for feature selection. *Neurocomputing* 172: 371–381.

15. Glover, F.W., and G.A. Kochenberger. 2006. *Handbook of metaheuristics*, vol. 57. Springer Science & Business Media.

16. Goldberg, D.E. 1989. *Genetic algorithms in search. Optimization, and machine learning*.

17. Hassan, M.M., Z. Uddin, A. Mohamed, and A. Almogren. 2018. A robust human activity recognition system using smartphone sensors and deep learning. *Future Generation Computer Systems* 81: 307–313.

18. He, Y.L., X.L. Zhang, W. Ao, and J.Z. Huang. 2018. Determining the optimal temperature parameter for softmax function in reinforcement learning. *Applied Soft Computing* 70: 80–85.

19. Hiram Ponce, S.G. 2019. An indoor predicting climate conditions approach using internet-of-things and artificial hydrocarbon networks. *Measurement* 135: 170–179.

20. Hou, M., H. Wang, Z. Xiao, and G. Zhang. 2018. An svm fall recognition algorithm based on a gravity acceleration sensor. *Systems Science & Control Engineering* 6 (3): 208–313.

21. Igual, R., C. Medrano, and I. Plaza. 2015. A comparison of public datasets for acceleration-based fall detection. *Medical Engineering & Physics* 37 (9): 870–878.

22. Jia, H., Z. Xing, and W. Song. 2019. Three dimensional pulse coupled neural network based on hybrid optimization algorithm for oil pollution image segmentation. *Remote Sensing* 11 (9): 1046.

23. Kennedy, J. 2010. *Particle swarm optimization. Encyclopedia of machine learning*, 760–766.

24. Klein, D. 2011. *Organic chemistry*. Wiley.

25. Kozina, S., H. Gjoreski, and M.G. Lustrek (2013). Efficient activity recognition and fall detection using accelerometers. In *International competition on evaluating AAL systems through competitive benchmarking*, 13–23. Springer.

26. Manne, P. 2016. Parallel particle swarm optimization. Master Thesis of North Dakota State University.

27. Marini, F., and B. Walczak. 2015. Particle swarm optimization (PSO). A tutorial. *Chemometrics and Intelligent Laboratory Systems* 149 (Part B): 153–165.

28. Martinez-Villasenor, L., H. Ponce, J. Brieva, E. Moya-Albor, J. Nunez-Martinez, and C. Penafort-Asturiano Up-fall detection dataset: A multimodal approach. *Sensors* X (X): XX–XX (in press).

29. Medrano, C., R. Igual, I. Plaza, and M. Castro. 2014. Detecting falls as novelties in acceleration patterns acquired with smartphones. *PloS One* 9 (4): e94811.

30. Mirjalili, S. 2015. How effective is the grey wolf optimizer in training multi-layer perceptrons. *Applied Intelligence* 150–161.

31. Mirjalili, S., S.M. Mirjalili, and A. Lewis. 2014. Grey wolf optimizer. *Advances in Engineering Software* 69: 46–61. https://doi.org/10.1016/j.advengsoft.2013.12.007. http://www.sciencedirect.com/science/article/pii/S0965997813001853.

32. Nayak, J., B. Naik, A. Jena, R.K. Barik, and H. Das. 2018. Nature inspired optimizations in cloud computing: applications and challenges. In *Cloud computing for optimization: Foundations, applications, and challenges*, 1–26. Springer.

33. Ofli, F., R. Chaudhry, G. Kurillo, R. Vidal, and R. Bajcsy. 2013. Berkeley MHAD: A comprehensive multimodal human action database. In *2013 IEEE workshop on applications of computer vision (WACV)*, 53–60. IEEE.

34. Ouyang, A., Z. Tang, X. Zhou, Y. Xu, G. Pan, and K. Li. 2015. Parallel hybrid PSO with CUDA for ID heat conduction equation. *Computers & Fluids* 110: 198–210.
35. Phinyomark, A., A. Nuidod, P. Phukpattaranont, and C. Limsakul. 2012. Feature extraction and reduction of wavelet transform coefficients for EMG pattern classification. *Elektronika ir Elektrotechnika* 122 (6): 27–32.
36. Ponce, H., Acevedo, M.: Design and equilibrium control of a force-balanced one-leg mechanism. In: Advances in Soft Computing, Lecture Notes in Computer Science, pp. 1–15. Springer (2018)
37. Ponce, H., and L. Martínez-Villasenor. 2017. Interpretability of artificial hydrocarbon networks for breast cancer classification. In *30th International joint conference on neural networks*, 3535–3542. IEEE.
38. Ponce, H., L. Martínez-Villasenor, and L. Miralles-Pechuán. 2016. A novel wearable sensor-based human activity recognition approach using artificial hydrocarbon networks. *Sensors* 16 (7): 1033.
39. Ponce, H., L. Miralles-Pechuán, L. Martínez-Villasenor. 2015. Artificial hydrocarbon networks for online sales prediction. In *Mexican international conference on artificial intelligence*, vol. 9414, 498–508. Springer.
40. Ponce, H., L. Miralles-Pechuán, and L. Martínez-Villasenor. 2016. A flexible approach for human activity recognition using artificial hydrocarbon networks. *Sensors* 16 (11): 1715.
41. Ponce, H., and P. Ponce. 2011. Artificial organic networks. In *Electronics, robotics and automotive mechanics conference (CERMA)*, 29–34. IEEE.
42. Ponce, H., P. Ponce, H. Bastida, and A. Molina. 2015. A novel robust liquid level controller for coupled-tanks system using artificial hydrocarbon networks. *Expert Systems With Applications* 42 (22): 8858–8867.
43. Ponce, H., P. Ponce, and A. Molina. 2013. Artificial hydrocarbon networks fuzzy inference system. *Mathematical Problems in Engineering* 2013 (531031): 1–13.
44. Ponce, H., P. Ponce, and A. Molina. 2014. Adaptive noise filtering based on artificial hydrocarbon networks: An application to audio signals. *Expert Systems With Applications* 41 (14): 6512–6523.
45. Ponce, H., P. Ponce, and A. Molina. 2014. *Artificial organic networks: Artificial intelligence based on carbon networks, Studies in Computational Intelligence*, vol. 521. Springer.
46. Ponce, H., P. Ponce, and A. Molina. 2015. The development of an artificial organic networks toolkit for labview. *Journal of Computational Chemistry* 36 (7): 478–492.
47. Ponce, P., H. Ponce, and A. Molina. 2017. Doubly fed induction generator (DFIG) wind turbine controlled by artificial organic networks. *Soft Computing* 1–13.
48. Precup, R.E., R.C. David, and E.M. Petriu. 2017. Grey wolf optimizer algorithm-based tuning of fuzzy control systems with reduced parametric sensitivity. *IEEE Transactions on Industrial Electronics* 64 (1): 527–534.
49. Preece, S.J., J.Y. Goulermas, L.P. Kenney, and D. Howard. 2009. A comparison of feature extraction methods for the classification of dynamic activities from accelerometer data. *IEEE Transactions on Biomedical Engineering* 56 (3): 871–879.
50. Rasekh, A., C.A. Chen, and Y. Lu. 2014. Human activity recognition using smartphone. arXiv preprint arXiv:1401.8212.
51. Sahani, R., C. Rout, J.C. Badajena, A.K. Jena, H. Das et al. 2018. Classification of intrusion detection using data mining techniques. In *Progress in Computing, Analytics and Networking*, 753–764. Springer.
52. Sebastian Gutierrez, H.P. 2019. An intelligent failure detection on a wireless sensor network for indoor climate conditions. *Sensors* 19 (4).
53. Talbi, N. 2019. Design of fuzzy controller rule base using bat algorithm. *Energy Procedia* 162: 241–250.
54. Teleimmersion Lab, U.O.C. 2013. Berkeley Multimodal Human Action Database (MHAD). http://tele-immersion.citris-uc.org/berkeley_mhad. Accessed 13 Dec 2018.
55. Vavoulas, G., M. Pediaditis, C. Chatzaki, E.G. Spanakis, and M. Tsiknakis. 2017. The mobifall dataset: Fall detection and classification with a smartphone. In *Artificial intelligence: Concepts, methodologies, tools, and applications*, 1218–1231. IGI Global.

56. Xu, G., and G. Yu. 2018. Reprint of: On convergence analysis of particle swarm optimization algorithm. *Journal of Computational and Applied Mathematics* 340: 709–717.
57. Xu, T., Y. Zhou, and J. Zhu. 2018. New advances and challenges of fall detection systems: A survey. *Applied Sciences* 8 (3): 418.
58. Yang, X.S., and A. Hossein Gandomi. 2012. Bat algorithm: A novel approach for global engineering optimization. *Engineering Computations* 29 (5): 464–483.
59. Zhang, B., W. Liu, S. Li, W. Wang, H. Zou, and Z. Dou. 2019. Short-term load forecasting based on wavelet neural network with adaptive mutation bat optimization algorithm. *IEEJ Transactions on Electrical and Electronic Engineering* 14 (3): 376–382.
60. Zhang, Y., S. Wang, and G. Ji. 2015. A comprehensive survey on particle swarm optimization algorithm and its applications. *Mathematical Problems in Engineering* 2015 (931256): 1–38.
61. Zhao, S., W. Li, W. Niu, R. Gravina, and G. Fortino. 2018. Recognition of human fall events based on single tri-axial gyroscope. In *2018 IEEE 15th International conference on networking, sensing and control (ICNSC)*, 1–6. IEEE.

Application of Genetic Algorithms for Unit Commitment and Economic Dispatch Problems in Microgrids

A. Rodríguez del Nozal, A. Tapia, L. Alvarado-Barrios and D. G. Reina

Abstract In the last decades, new types of generation technologies have emerged and have been gradually integrated into the existing power systems, moving their classical architectures to distributed systems. In spite of the positive features associated to this paradigm, new problems arise such as coordination and uncertainty. In this framework, microgrids (MGs) constitute an effective solution to deal with the coordination and operation of these distributed energy resources. This book chapter proposes a Genetic Algorithm (GA) to address the combined problem of Unit Commitment (UC) and Economic Dispatch (ED). With this end, a detailed model of a MG is introduced together with all the control variables and power restrictions. In order to optimally operate the MG, two operation modes are introduced, which attend to optimize economical factors and the robustness of the solution with respect power demand uncertainty. Therefore, it achieves a robust design that guarantees the power supply for different confidence levels. Finally, the algorithm is applied to an example scenario to illustrate its performance.

Keywords Microgrids · Unit commitment · Economic dispatch · Genetic algorithm

A. Rodríguez del Nozal
Department of Electrical Engineering, Universidad de Sevilla, Seville, Spain
e-mail: arnozal@us.es

A. Tapia · L. Alvarado-Barrios
Departamento de Ingeniería, Universidad Loyola Andalucía, Seville, Spain
e-mail: atapia@uloyola.es

L. Alvarado-Barrios
e-mail: lalvarado@uloyola.es

D. G. Reina (✉)
Electronic Engineering Department, University of Seville, Seville, Spain
e-mail: d.gutierrez.reina@gmail.com

© Springer Nature Switzerland AG 2020
M. Rout et al. (eds.), *Nature Inspired Computing for Data Science*,
Studies in Computational Intelligence 871,
https://doi.org/10.1007/978-3-030-33820-6_6

1 Introduction

Nowadays, the centralized electric sector is gradually being transformed into a distributed scheme, strongly characterized by the increasing integration of Renewable Energy Sources (RES) [1, 2]. This process introduces a new paradigm in the future of the sector, in pursuit of the decarbonization, the depletion of the energy dependency on third countries and the reduction of transport, and distribution costs of the actual systems, among others [3]. The most interesting characteristic of the distributed scheme lies in the ease of integration of numerous small distributed RES. The study of these systems has become an area of particular interest, constituting a social and financial object of investment by governments and energy companies [4]. In this context of transformation into the smart grids of the future, micro-grids play a fundamental role [5–7].

The use of RES in the electric sector is gaining relevance (RES actually cover around 20% of the total power generation [8]), thus, the need of Energy Storage Systems (ESS) to deal with the stochastic nature of these sources becomes essential [9]. These systems can provide energy balance [10], improve the service quality [11], add additional functionalities (such as inertial response, spinning reserve, etc) [12, 13] and provide additional services to the final customer [14].

A possible tool to overcome these problems is the microgrid. Although the concept of microgrid (MG) [15] has been pervasively used in the literature since its introduction [16, 17], its definition is still a matter of discussion. Nevertheless, it can be described as a set of loads, Distributed Generation (DG) units and EES that operate in coordination to supply energy, either isolated or connected to a host power system at the distribution level at a single point of connection, defined as the Point of Common Coupling (PCC). On the basis of this definition, MGs are considered as low and medium-voltage distribution grids [18]. It is relevant to note that MGs can have an arbitrary scheme, generally including the combination of conventional generation units (such as diesel engines (DE), micro-turbines (MT), fuel-cells (FC), or combined heat and power plants), RES (such as solar photo-voltaic (PV) panels or wind turbines (WT)) and ESS [6], operating as an independent controllable unit [4, 19].

The uncertainty derived of the stochastic nature of RES, such as solar and wind energy, strongly affects the performance and conditions the reliability of the distribution grids [20, 21]. This issue, together with the voltage increase that penetration of distributed generation units imply, motivate the use of EES. These systems provide the capability of deciding the optimal operation scheme of the MG, by working either isolated or connected to the grid [22]. ESS are especially relevant in isolated MGs, in order to assure the correct balance between generation and load consumption [23]. ESS can also reduce the operation cost by storing energy during low cost periods [24], and can also be used to reduce the load peak, allowing to smooth the demand curve [25]. Among the diversity of ESS available, the Battery Energy Storage Systems (BESS) have gained especial interest in the field, given their clear advantages such as fast response or control capacity [26, 27]. In addition, BESS also permit a wide variety of applications, ranging from improving the energy quality to perform long-term energy management, and covering additional aspects such as the improvement of the reliability and the transmission [28, 29].

The Energy Management Systems (EMS) constitute the elements that are responsible of controlling the MGs [30]. These systems are capable of operating in the most suitable way the set of generation units, dispatching the active energy from RES and controlling the load consumption, following an economic criterion [31], in coordination with the Distribution System Operator (DSO) [32]. Nevertheless, given the stochastic nature of RES, together with their increasing protagonism in the distributed schemes, finding the most adequate management strategy of the MGs becomes a challenge itself [33, 34], which is especially intensified by the inclusion of ESS. Not only efficiency is directly related to this management but also safety and privacy are involved, what makes the Unit Commitment (UC) problem a useful tool to study the optimal power distribution through the MG [35]. It contributes to the improvement of the stability and reliability of the system, while considering the environmental and economic impact associated to its operation [36].

1.1 The Unit Commitment Problem and the Proposed GA-Based Approach

The Unit Commitment (UC) problem [37] consists of determining the scheduling of the generation units with the purpose of satisfying the power demand. Once the UC problem has been solved, the Economic Dispatch (EC) is responsible of assigning the generation levels to the programmed generation units, in such way that the resources are used in the most efficient way, and satisfying the physical constraints imposed by the generating units (power balance, power limits, etc) and the grid (power transfer limits, spinning reserve, etc). These problems are solved for a certain period of time, and attending to specific constraints, generally related to a load forecast, spinning reserve, and electricity market context [38]. In the context of distributed MG, it is worth pointing out the increase of complexity [39] that the RES stochasticity and the management of ESS imply, resulting in a non-convex mathematical problem, which combines strong non-linearities and mixed-integer formulation.

Although the simplification of this problem by a model linearization is frequent, it might introduce inaccuracies in the results. To deal with this, in this work a meta-heuristic approach is proposed. Alike lineal programming based approaches, meta-heuristic algorithms allow to deal with non-linearities in an effective way, since the search engine is based on genetic operations (selection, crossover, and mutation) among potential solutions, making them ideal for addressing the UC problem. The proposed GA-based approach encodes potential solutions of the UC problem in a chromosome-like structure. Such solutions include the power provided in a time period of 24 h by the elements that are costlier in a microgrid such as the micro-turbine and the diesel engine since the power supplied by the ESS can be derived from the power balance (see more details in following sections). Regarding the RES, they will always contribute as the climate conditions are favorable. The constraints of the UC problem in terms of maximum power provided, power demand and balance, energy store systems, and spinning reserve are included in the power by penalizing invalid solutions through death penalty. Therefore, discarding bad solution by not allowing them to participate in the genetic operations (more details on the GA implementation can be found in Sect. 5).

The main contribution of this work are:

1. Improving the scheduling problem of MGs by considering different levels of reliability according to the uncertainty of the forecast demand. Thus, the spinning reserve of dispatchable generation units guarantees that the variability of the demand is always covered.
2. The introduction of two main modes of operation: a cost effective operation mode, in which the main aim is to minimize the operation and maintenance cost of the overall system; and robust operation mode, which pursuits the same objectives while considering an additional level of reliability.
3. The application of evolutionary algorithms to solve the combined problem.

This chapter is organized as follows: In Sect. 1.2, a brief summary with the main contributions in the literature is presented, as well as an introduction to the motivations and objectives of this work. In Sect. 2, the problem is presented, being the concept of MG introduced. A model of the problem is developed in Sect. 3, being the different elements of the studied MG explained in depth. In Sect. 4 the operation modes of the MG are presented. The proposed computational evolutionary approach is developed in Sect. 5. This approach is then applied to proposed MG, being the results presented and studied in Sect. 6. Finally, the main conclusions of this work are summarized in Sect. 7.

1.2   Related Work

A relevant review on the objectives, constraints and algorithms used in EMS for MGs can be found in [40]. The authors provide an extensive summary of the main strategies used to address this problem, and provide a wide overview of the state of the art in the UC problem for MGs. In addition, a classification of the main techniques applied is provided.

The UC problem is usually classified as Mixed-Integer Non-Linear Programming (MINLP) [41–43]. A relevant work can be found in [41], where authors define a profit-based UC problem with non-linear emissions penalty, obtained through a mixed-integer formulation. The authors address the non-linearities of the fuel cost function by means of a piece-wise linearization. This approach is applied to different schemes, demonstrating its capability of obtaining good results. In [42] a similar strategy is proposed, demonstrating a high robustness. Regarding Linear Programming (LP), the authors in [44] develop a MILP to determine a fair cost distribution among smart homes with MG, resulting in noticeable cost reductions and fair cost distribution among multiple homes and under different scenarios.

Despite MILP and MINLP approaches provide a high efficiency, it is relevant to note that the computational cost increases exponentially with the number of integer variables and thus these approaches are not suitable for bigger grids. Nevertheless, several strategies have been proposed in the literature to deal with this drawback, such as using a multi-step method to consider the combinatorial calculations of unit outage events [45], solving separately the different scales [46] or other improved strategies to permit MILP to address the UC problem in big scale grids [47]. Regarding this issue, an interesting work is found in [48], where the authors propose the decomposition of the problem into UC and Optimal Power Flow (OPF) problems, and thus the original MINLP problem is transformed into two separated problems, MILP and a NLP, respectively. Also, Mixex-Integer Quadratic Problem (MIQP) formulation has also been proposed to address the UC problem. A relevant work can be found in [49]. In this work, an improved MIQP is proposed to increase its efficiency by means of relaxation and decoupling techniques.

Apart from MILP, MINLP and MIQP approaches, a wide variety of optimization algorithms have been proposed in the literature to address the UC problem in MGs [50, 51]. Although among all these works a broad selection of architectures can be found, these generally include controllable RES, operating either isolated or connected to the grid. The use of evolutionary algorithms in order to tackle this problem looks a promising strategy based on the results obtained in other problems with similar complexity. For instance, consider [52] where GA are used to classify the diabetes mellitus disease. In [53] a classification of intrusion detection using data mining techniques is presented. Finally, in [54] introduces the topic of information security in biomedical signal processing.

In [36] the authors develop two optimization algorithms on the basis of a Genetic Algorithm (GA) and MILP, respectively, being them applied to an European radial low-voltage MG, for which the authors define a set of objective functions and constraints, attending to different modes of operation of the MG (cost-efficiency, grid supporting, reliable isolated operation, eco-friendliness and multi-functional operations). The results prove the GA to provide a better performance in comparison to MILP. In a similar way, in [50] the start-up and shut-down costs associated to the distributed generators are considered resulting in a more complex problem which is addressed by means of a hybrid optimal algorithm. This consists of decomposing the problem into Integer Programming (IP) to solve the UC problem and Non Linear Programming (NLP) to solve the ED problem. After that, both GA and interior-point algorithms are used to solve the IP and NLP problem, respectively. In [55], the authors propose an Enhanced Genetic Algorithm (EGA) to address both the UC and ED problems simultaneously. The goodness of this strategy is proven by a set of case-studies applied to a test MG. A good performance is observed in the simulation results of the optimizer under grid-connected operation mode, being the required computational resources strongly increased for the isolated mode. Another work of relevance can be found in [56], where a Memory-Based Genetic Algorithm (MGA) is proposed to address the problem of minimizing the overall energy cost. The results are compared with two variants of Particle Swarm Optimization (PSO) algorithms, showing a superior behavior. In a similar way, in [57] the authors address the same problem by improving a GA by integrating a Simulated Annealing (SA) algorithm to accelerate the convergence, demonstrating a clear advantage in convergence time.

Regarding the design of ESS, authors in [58] propose a fuzzy multi-objective optimization model to optimally size the ESS. The proposed approach lies in a Chaotic Optimization Algorithm (CGA), introduced into a Binary particles Swarm Optimization (BPSO) algorithm, resulting in a Chaotic BPSO (CBPSO). Additionally, some authors propose the use of Monte Carlo simulations as a effective tool to generate scenarios. For example, in [59] the authors implement neural network based prediction intervals (PIs) to quantify forecast uncertainty of wind power. A GA is then used to solve the Stochastic Unit Commitment (SUC) problem, while the Monte Carlo simulations are used to create the scenarios. The results demonstrate a better performance of the CBPSO in comparison with the original BPSO. In a similar way, the authors in [51] consider the uncertainty of demand and wind source, modeling it by means of absolute percentage error. Monte Carlo simulations are used to generate demand and wind power scenarios. To optimize the UC problem, the priority list method (PL algorithm) [60] is applied.

The main contributions of this work with respect to the ones presented in this section are:

- A cost-effective operation mode is considered in which the main objective is minimizing the operation and maintenance costs of the system.
- Different levels of reliability, in accordance with the uncertainty of the forecast demand, are considered to improve the scheduling problem of the MG. The spinning reserve of the dispatchable units guarantees the coverage of the demand variability.

2 Problem Description

The electric power system is defined as a network of electrical components deployed to supply, transfer, store, and use electric power. Traditional power systems have a top-down operated architecture in which their components can be broadly classified as generators, that supply the electric power; the transmission system, that carries the power to the load centers; and the distribution system, that feeds the loads. This paradigm is now being altered by the integration of the so-called *prosumers*, that is, agents at the electricity grid that can both consume and produce power. This new situation, on the one hand, creates new opportunities for a better network regulation and penetration of renewable energy and, on the other hand, introduces new challenges in order to meet the demanded power and the generation in the network.

The inclusion of these renewable resources in the electrical grid has many associated problems that must be deeply studied [61]. Nevertheless, this challenge can be

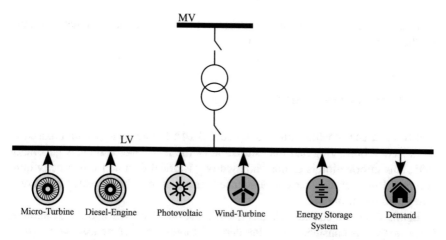

Fig. 1 Structure of the MG considered

partially addressed by MGs. In a MG the different DER are controlled in a consistently more decentralized way, thereby reducing the control burden on the grid and permitting them to provide their full benefits [62].

Several strategies can be applied to meet the required power demand in a MG [35]. This problem, which results of the combination of the Unit Commitment (UC) and Economic Dispatch (ED) problems, aims to minimize a certain cost function, while satisfying all of the constraints so that a given security level can be met [37, 63, 64]. On the one hand, the UC problem studies which generators must be in operation at each time instant t meanwhile, on the other hand, the ED studies, once known the generators connected to the grid, the power that must be supplied by each of them.

This work considers a MG whose scheme is depicted in Fig. 1. Note that this system is composed by two renewable resources: a wind turbine (WT) and a photovoltaic plant (PV); and two dispatchable units: a diesel engine (DE) and a micro-turbine (MT). The term dispatchable means that the power given by those units can be modified arbitrarily (just fulfilling the physical restrictions of the generators). In addition, an ESS is also considered in order to give resilience to the overall MG. Thus, the problem studied is to determine the power supplied by the different generators as well as the power supplied or demanded by the ESS at every time instant, minimizing certain cost function. This cost function may vary according to the operation mode fixed. All these issues will be tackled in the following sections.

3 System Modeling

The following points introduce the model adopted for each of the elements considered in the MG. All the models presented in this section have been used widely, however, in practice, their accuracy can vary substantially. These inaccuracies can be taken into consideration when solving the problem in order to guarantee a reliable solution for the problem.

3.1 Demand Forecasting

The UC and ED problems are usually solved 24 h before its application based on the demand forecasting and the estimation of the renewable resources generation [65]. This information is estimated based on historical data and it is usually quite accurate. However, they have an associated error that must be corrected in the real-time operation of the MG.

In this work, we consider an accurate estimation of the WT and PV generation. Conversely, it is considered that the demand forecasting has an associate error at every time instant t that follows a normal distribution, $\mathcal{N}(\mu_t, \sigma_t)$, whose average, μ_t, is considered zero and the standard deviation, σ_t, varies according to the uncertainty taken into consideration. Thus, if a high value of σ_t is considered, the uncertainty in the demand forecasting will be greater and a more robust solution must be contributed in order to guarantee the electrical supply under normal operation.

Let us define $P_{DM,t}$ as the real power demanded from the MG at time t and let $\tilde{P}_{DM,t}$ be the one day ahead prediction of that demand. Thus, it is clear that variables can be related as:

$$\tilde{P}_{DM,t} + e_{DM,t} = P_{DM,t}, \quad \forall t, \tag{1}$$

where $e_{DM,t} \sim \mathcal{N}(0, \sigma_t)$ is the estimation error.

It is well-known that, when considering a normal distribution $\mathcal{N}(\mu_t, \sigma_t)$, the values less than one standard deviation, σ_t, away from the mean, μ_t, account for 68.27% of the set, while two and three standard deviation away from the mean account for 95.45% and 99.73% of the set, respectively [66]. Thus, if the solution of the problem guarantees the supply of a power demanded $P_{DM,t} \in [\tilde{P}_{DM,t} - 3\sigma_t, \tilde{P}_{DM,t} + 3\sigma_t]$, then, a reliable solution is achieved with a high confidence interval. To obtain a good estimation of the demand forecasting uncertainties, an ARMA (Auto Regressive Moving Average) model was trained with two years of estimation data.

3.2 Photo-Voltaic Generator

A PV system, is a power system designed to supply usable solar power by means of photovoltaics. It is composed of several solar panels that absorb the sunlight energy and transform it into DC current. By using a solar inverter, the DC current is converted into AC current that is finally injected into the main grid.

The power produced by this generator depends on many factors. Most of them attend to climate conditions such as the solar irradiation or the cells temperature, but there exist other factors that depend on the characteristics of the module. For example, the module degradation. On the other hand, the power generated by these kind of systems is directly related with the operational point of the solar modules. This operational point is fixed by the power inverter by using a Maximum Power Point Tracking (MPPT). There exists a large number of strategies in order to find the Maximum Power Point (MPP), nevertheless, that is not the focus of the present work.

In this work we consider the simplified model presented in [67]. This model computes the power given by the PV system from the comparison of the real-time state of the module and the state under the Standard Test Conditions (STC). These parameters are easily accessible from the data sheet of the modules and correspond to the parameters obtained after testing the PV module in the laboratory with a cell temperature of 25 °C and an irradiance of 1000 W/m² with an air mass 1.5.

In particular the model considered is exposed next:

$$P_{PV,t} = P_{STC} \frac{n \cdot E_{M,t}}{E_{STC}} [1 + k(T_{M,t} - T_{STC})], \quad \forall t, \qquad (2)$$

where $P_{PV,t}$ denotes the output power of the PV plant at time t, $E_{M,t}$ is the solar irradiance at t and P_{STC}, E_{STC} and T_{STC} are the maximum power, the irradiance and the temperature under Standar Test Conditions (STC), respectively. Finally, n denotes the number of PV panels, k the power temperature coefficient (%/°C) and $T_{M,t}$ is the temperature of the module at time t, which can be calculated as $T_{M,t} = T_{amb} + \varepsilon_{PV} \frac{E_{M,t}}{E_{STC}}$, where T_{amb} is the ambient temperature (°C) and ε_{PV} is a constant module provided by the manufacturer.

Note that this model does not take under consideration other factors as the module degradation or the error derived from using a inefficient MPPT algorithm.

3.3 Wind Turbine

Analogously to the PV generator model, there exist a high number of publications dealing with the modeling and simulation of wind turbines [68, 69]. Most of these models have a great level of detail and are later used to perform stability analysis of power networks. In this chapter we make use of the following simplified version of the wind turbine model:

$$P_{WT,t} = \begin{cases} 0 & \text{for} \quad v < v_{ci}, \\ P_r \frac{v^3 - v_{ci}^3}{v_r^3 - v_{ci}^3} & \text{for } v_{ci} < v < v_r, \\ P_r & \text{for } v_r < v < v_{co}, \\ 0 & \text{for} \quad v > v_{co}, \end{cases} \quad \forall t, \qquad (3)$$

where P_r and v_r represent the rated power (kW) and wind speed (m/s), respectively, and v_{ci}, v_{co} and v are the cut-in, cut-out and actual wind speed, respectively. A more detailed information about this model can be found in [70].

3.4 Diesel Engine

The most common type of generator used in MGs is the diesel engine. The diesel engine plays an important role in the problem, especially when it comes to small-scale MGs. This fact is a consequence of the inconsistency of renewable weighted against diesel technology's durability. The wind does not blow consistently nor does the sun shine on a reliable timetable across much of the country, and when these renewables are available may not be when power is needed.

The operation cost of a this dispatchable unit can be expressed as a function of its real power output and it is usually modeled as a quadratic polynomial with the shape:

$$C_{DE,t} = d_{DE} + e_{DE} P_{DE,t} + f_{DE} P_{DE,t}^2, \quad \forall t, \tag{4}$$

where $C_{DE,t}$ is the total fuel cost (€/h) at time t, $P_{DE,t}$ is the power output at time t and d_{DE} (€/h), e_{DE} (€/kW h) and f_{DE} (€/kW^2 h) are parameters specified by the manufacturer [71], and are strongly conditioned by the type of engine, generator and fuel. Additionally, let P_{DE}^{min} and P_{DE}^{max} be the minimum and maximum power that the DE can supply at every time instant. Then:

$$P_{DE}^{min} \leq P_{DE,t} \leq P_{DE}^{max}. \tag{5}$$

Recall that, as it was done in the rest of generators, there exist a numerous variety of models supplying the operational cost of the unit. However, we consider this model due to the fact that is one of the most commonly used.

3.5 Microturbine

A turbine is a device that converts the flow of a fluid into mechanical motion for generation of electricity. The prefix "micro" implies that is a small device. Thus, the MT model is similar to the DE model and it is also considered as a dispatchable resource.

However, in this case the operation cost curve parameters are adopted in order to model the performance and efficiency of a MT unit [72]. Thus, the operation cost can be expressed as:

$$C_{MT,t} = d_{MT} + e_{MT} P_{DE,t} + f_{MT} P_{MT,t}^2, \quad \forall t, \tag{6}$$

where $C_{MT,t}$ is the total gas cost (€/h) at time t, $P_{MT,t}$ is the power output at time t and $d_{MT}(e/h)$, e_{MT} (e/Kw h) and f_{MT} (e/Kw^2 h) are parameters specified by the manufacturer [71, 72], and depend on the type of turbine, generator and gas.

Analogously with the DE, the power generated by the MT is physically limited:

$$P_{MT}^{min} \leq P_{MT,t} \leq P_{MT}^{max}, \tag{7}$$

where P_{MT}^{min} and P_{MT}^{max} are the minimum and maximum power that the DE can supply at every time instant.

3.6 Spinning Reserve

The term spinning reserve is widely used in the literature [73, 74] but its definition varies leading to confusion. In this work the following definition for the spinning reserve is considered [75]: the spinning reserve is the unused capacity that can be activated on decision of the system operator and is provided by devices, which are synchronized to the network and able to change the active power.

Note that this capacity is only found in the dispatchable units, as they are the only units that can be used on demand. Thus, we denote $R_{DE,t}$ and $R_{MT,t}$ the spinning reserve of the DE and the MT at time instant t, respectively:

$$R_{DE,t} = P_{DE}^{max} - P_{DE,t}, \qquad R_{MT,t} = P_{MT}^{max} - P_{MT,t}, \tag{8}$$

3.7 Energy Storage System

Batteries are electro-chemical devices that store energy from other AC or DC sources for later use [72]. Due to the uncertain character of the wind and PV power generation (that depends on the meteorological conditions), the battery is a key component of a MG. The battery allows a MG to cover demand peaks and to store excesses of generation due to its quick response.

Let us consider a set of batteries that conforms an ESS that is particularized by three parameters:

1. The capacity: is the amount of electric charge that the ESS can deliver at the rated voltage.
2. The maximum charge rate: is a measure of the rate at which a battery can be charged.
3. The maximum discharge rate: is a measure of the rate at which a battery can be discharged.

We denote $P_{ESS,t}$ as the power demanded or supplied by the ESS at time t. The value of $P_{ESS,t}$ can be both positive or negative. If $P_{ESS,t} > 0$ means that the ESS is supplying power to the MG meanwhile when $P_{ESS,t} < 0$ the ESS is charging and acting as a consumer. Attending to the charging/discharging rate, both parameters are bounded due to physical limitations:

$$P_{ESS,t} \leq P_{ESS}^{max}, \tag{9}$$

$$P_{ESS,t} \geq P_{ESS}^{min}. \tag{10}$$

Let SOC_t be the State of Charge of the ESS at every time t. According to the energy stored in the battery, the SOC_t will take different values. The SOC_t is lower and upper bounded according to the capacity of the battery and the minimum admissible state of charge. That is:

$$SOC^{min} \leq SOC_t \leq SOC^{max}, \quad \forall t, \tag{11}$$

where SOC^{min} and SOC^{max} are the lower and upper boundaries for the SOC_t. Additionally, recall that a battery is a dynamical system that evolves with time:

$$SOC_{t+1} = SOC_t - \begin{cases} P_{ESS,t}\Delta t\eta_c & \text{for } P_{ESS,t} < 0 \\ \frac{P_{ESS,t}\Delta t}{\eta_d} & \text{for } P_{ESS,t} > 0 \end{cases}, \quad \forall t, \tag{12}$$

where η_c and η_d are respectively the charging and discharging efficiency and Δt is the time between samples. It is important to highlight that we have adopted the convention in which if $P_{ESS,t} > 0$, the ESS is discharging and consequently the SOC_{t+1} decreases. Conversely, when $P_{ESS,t} < 0$ and ESS is charging, the SOC_{t+1} increases its value.

As it is exposed in [76], different models of batteries can be found in the literature with different degrees of complexity and simulation quality. In the present work we have used a simplified model of the battery that takes into consideration the main characteristics of it.

3.8 Power Balance

Several studies about the UC and ED problems can be found in the literature (see for instance [77] or [78]). Some of them consider a MG connected to the main distribution grid as it was done in [36]. In contrast with these approaches we will consider an isolated MG, i.e. it lacks connection to the distribution network. This fact implies that the power demanded by the customers must be locally supplied by the DER presented in the MG. Thus, the robustness in the UC and ED solution has a vital importance for the correct functioning of the microgrid.

Thus, the power balance must be fulfilled at every time interval t considered:

$$P_{DM,t} = P_{PV,t} + P_{WT,t} + P_{DE,t} + P_{MT,t} + P_{ESS,t}, \quad \forall t. \tag{13}$$

It is worth highlighting that, when considering the power balance, it must be taken into consideration possible uncertainties in the demand forecasting in order to guarantee the reliability of the solution provided. This reliability is met with the well-sizing of the spinning reserve. Thus, the following constraint must be added to the problem:

$$R_{DE,t} + R_{MT,t} \geq n_r \sigma_t, \qquad (14)$$

where σ_t is the standard deviation of the estimation error and n_r is a scalar parameter that, once fixed, establishes the reliability degree of the solution adopted. A deeper study of this issue is presented in Sect. 4.2.

3.9 Other Operation Costs

In addition to the aforementioned operation costs, there exist a few extra costs associated with the MG functioning. We will consider the followings:

1. Maintenance cost: The operating and maintenance cost of each dispatchable generating unit is assumed to be proportional to the power production [72, 79]:

$$C_{OM_{DE}} = k_{OM_{DE}} \sum_{\forall t} P_{DE,t} \Delta t, \quad \forall t, \qquad (15)$$

$$C_{OM_{MT}} = k_{OM_{MT}} \sum_{\forall t} P_{MT,t} \Delta t, \quad \forall t, \qquad (16)$$

where $K_{OM_{DE}}$ (€/kWh) and $K_{OM_{MT}}$ (€/kWh) are the maintenance costs of the DE and the MT, respectively.

2. Start-up cost: The generator start-up cost depends on the time at which the unit has been off prior to start up [80]. Thus, the start-up cost at any given time can be estimated as [81]:

$$C_{SU_{DE}} = a_{DE} + b_{DE} \left[1 - exp \left(\frac{-T_{DE,OFF}}{c_{DE}} \right) \right], \qquad (17)$$

$$C_{SU_{MT}} = a_{MT} + b_{MT} \left[1 - exp \left(\frac{-T_{MT,OFF}}{c_{MT}} \right) \right], \qquad (18)$$

where a_{DE}, a_{MT} (€) are the hot start-up costs, b_{DE}, b_{MT} (€) are the cold start-up costs, c_{DE}, c_{MT} (€) are the unit cooling time constant and $T_{DE,OFF}$ and $T_{MT,OFF}$ represent the time that each unit has been off (for a more detailed explanation about this expression, reader is referred to [81]).

4 MG Operation Modes

Previous sections have introduced all the restrictions of the problem according to physical limitations of the generators. In this section, the modes of operation of the MG are first presented.

4.1 Cost-Effective Operation Mode

This operation mode is focused on the reduction of the overall cost of the MG without taking into account the possible deviations in the demand forecasting. Thus, the cost function considers the sequel parameters:

- Cost of operation and maintenance of the dispatchable units (DE and MT).
- Start-up costs of the dispatchable units (DE and MT).
- Fuel and gas costs of the DE and MT, respectively.

Thereby, the cost function is mathematically defined as follows:

$$CF_{ec} = C_{OM_{DE}} + C_{OM_{MT}} + C_{SU_{DE}} + C_{SU_{MT}} + \sum_{\forall t} \left(C_{DE,t} + C_{MT,t}\right). \quad (19)$$

4.2 Robust Operation Mode

The consideration of uncertainties in the demand forecasting introduces a complication in the UC and ED problem. This problem has been classically denoted as Stochastic Unit Commitment (SUC). The SUC problem has attracted the attention of many researchers in the area that have tackled the problem from several perspectives [77].

In this work we try to provide a robust solution to the problem. To do that, it looks reasonable to design the spinning reserve in order to cover the possible demand deviations derived from an inefficient prediction. Thus, we can fix different values of the parameter n_r, first presented in (14), to establish different levels of reliability of the problem (Fig. 2 shows the reliability level for values of $n_r = 1$, $n_r = 2$ and $n_r = 3$).

Note that this operation mode is complementary to the cost-effective operation mode since it includes additional constraints to the problem and preserve the cost function expression (19). In this mode, more expensive solutions are expected to be obtained, given the severe constraints related to robustness that they must fulfill.

Fig. 2 Normal distribution of the demand estimation error

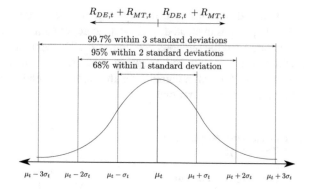

5 Evolutionary Computational Approach

Genetic algorithms (GAs) are meta-heuristic algorithms that have been widely used for solving complex engineering problems [82, 83]. GAs are based on the Darwinian theory of biological evolution. The basic idea is to have a set of potential solutions, that is called the population, which evolves over a number of generations by using genetic operators such as selection, crossover, and mutation [84]. The potential solutions encode in a chromosome-like structure the designing variables. This structure is called an individual and each designing variable represents a gene. In this work, the designing variables are the power supplied/consumed by the MT, DE, and the ESS since they are the elements of the considered MG that have higher impact on the operational cost of the MG (see Fig. 1). RES, such as PV and WT, will always contribute to the demanded power as long as the solar and wind conditions are both favorable, since their maintenance and operational costs can be neglected with respect to the DE and MT generators. Furthermore, since the power balance at each time instant should be fulfilled, whether the ESS supply or consume power at each time interval t, is determined by the power balance between the demand and the generation of the rest of elements of the MG.

5.1 Genetic Algorithm Implementation

There are many possible implementations of single-objective GAs [85]. In the proposed approach, a mupluslambda variant has been used [86]. Algorithm 1 shows the implemented approach. The algorithm begins with a random initial population P_i, which is evaluated. It is important that the individuals that form the initial population are valid. This aspect is relevant in problems with multiple constrains as the one described in this paper (see Sect. 3). Then, the offspring μ is created by using crossover and mutation operations. p_{cx} and p_{mut} refer to the crossover and mutation probability respectively. Both p_{cx} and p_{mut} are hyper-parameters of the algorithm that should be selected carefully to guarantee a good converge of the GA. Next, the offspring is evaluated, and the new population λ is selected from the offspring generated and the previous population P_g. This approach guarantees a good level of elitism since parents and offspring compete each other to be selected for the next generation [86]. After completion, the algorithm returns the best individual obtained throughout the generations.

```
1  Create initial population P_i;
2  Evaluate P_i;
3  P_g = P_i;
4  while stop == False do
5  |    Parents' selection;
6  |    Create offspring μ (crossover p_cx and mutation p_mut);
7  |    Evaluate μ ;
8  |    Select new population λ (μ + P_g);
9  |    P_g = λ;
10 end
```
Algorithm 1: GA mupluslambda

5.1.1 Individual Representation

Each individual represents a list containing the power supplied by both the MT and DE for a time period of 24 h, using intervals of 1 h. Therefore, the size of each individual is 48 genes. It is worth recalling that the power supplied or consumed by the ESS can be derived from the power balance, and also, the renewable sources will always contribute to the demanded power as long as the climate conditions are suitable. Figure 3 shows the encoding used to represents the individuals of the GA.

5.1.2 Fitness Function

The fitness function used depends on the operation mode of the microgrid. Two operation modes have been defined in Sect. 4 that use a single-objective fitness function, such as cost-effective and robust operation modes.

In the case of cost-effective mode, the objective function is defined as:

$$\begin{cases} if \text{ \textbf{solution valid} } F = \text{Eq. (19)}, \\ \quad else \quad\quad F = \infty, \quad according\ to \text{ Eqs. (5), (7), (9)–(14).} \end{cases} \quad (20)$$

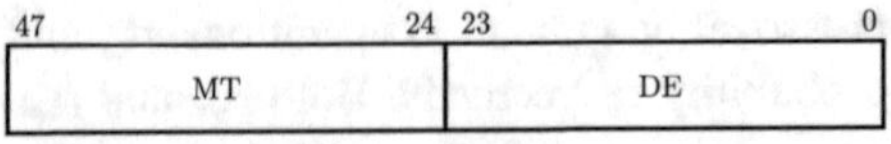

Fig. 3 Individual representation

47	24	23	0
MT		DE	

In the case of robust mode, the fitness function used is:

$$\begin{cases} if \ \textbf{solution valid} \ F = \text{Eq. (19)}, \\ \quad\quad else \quad\quad F = \infty, \quad according \ to \ \text{Eqs. (5), (7), (8)–(14)}. \end{cases} \quad (21)$$

It is important to highlight that in both fitness functions (20) and (21), the death penalty has been used to penalize invalid solutions. Therefore, these invalid solutions will not participate in the genetic operations since they will not be chosen by the selection scheme.

5.1.3 Genetic Operators

Tournament selection mechanism has been used since it provides suitable results [85]. In each tournament, a number of individuals are randomly selected, which compete each other to be chosen as a parent; the best one is then selected as one of the parents to be used in crossover and mutation operations [85]. A tournament size of three has been demonstrated to be suitable for the majority of problems. Regarding the crossover operation, two schemes have been evaluated such as two-point and Simulated Binary Crossover (SBX) [87] methods. The two-point crossover consists of swapping the genetic information of two parents using two points as the indexes of the genetic exchange. SBX is a method to simulate one point crossover for continuous variables. An eta hyper-parameter is used to determine the similarity among selected parents and children. A high eta value will produce children resembling to their parents, while a small eta value will produce solutions much more different. The mutation scheme used is a tailored Gaussian mutation algorithm, where each variable can change according to a Gaussian distribution with mean μ and standard deviation σ. The variability of a given gene after mutation depends on the value of σ.

6 Simulation Results

This section presents the main results of the paper. First, a particularization of the MG presented in Sect. 2 is presented. After that, the GA previously introduced is applied to different case-studies. Finally, the results are shown together with the values of the cost functions for the different cases. A simulation framework has been developed in Python using DEAP library [88]. The simulator is available in [89].

6.1 Example Scenario Settings

As it was introduced in Sect. 2, the following components are considered in the MG:

- Disel Engine (dispatchable unit).
- Micro-Turbine (dispatchable unit).
- PV system.
- Wind Turbine.
- Energy Storage System.

The problem is solved considering a sample time of one hour and a time window of 24 h. Note that these parameters can be altered affecting the size of the variable vector of the problem but preserving the formulation adopted through the chapter.

The configuration parameters of each of the dispatchable generators together with their operation, start-up and maintenance costs coefficients are exposed in Table 1.

Figure 4 shows the operation costs of the dispatchable units are shown in Fig. 4. Recall that when the amount of energy provided by a dispatchable unit is lower than 55 kW the DE is the most economical solution. However, when the required energy increases, the MT is more adequate in terms of cost.

The start-up cost is also a main factor when solving the problem. Figure 5 shows the start-up cost for both dispatchable units with respect to the time that the unit has been off. As it can be appreciate, the cold start of the MT has a higher cost than the DE, but, after five hours of not operation, the MT is most economical solution.

Table 1 Characteristic parameters of each of the generators considered in the MG

i	P_i^{min} (kW)	P_i^{max} (kW)	d_i (€/h)	e_i (€/kWh)	f_i (€/kW² h)
DE	5	80	1.9250	0.2455	0.0012
MT	20	140	7.4344	0.2015	0.0002
i	a_i (€)	b_i (€)	c_i (€)	K_{OM_i} (€/kWh)	
DE	0.3	0.4	5.2	0.01258	
MT	0.4	0.28	7.1	0.00587	

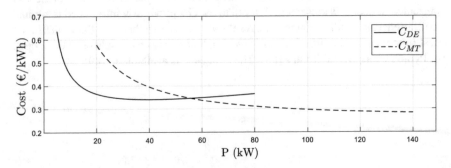

Fig. 4 Cost of operation of the DE and MT in terms of the power supply

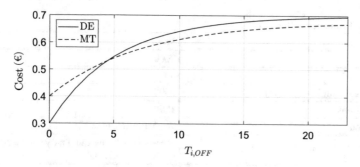

Fig. 5 Start-up cost of the dispatchable units in terms of the time the unit has been off

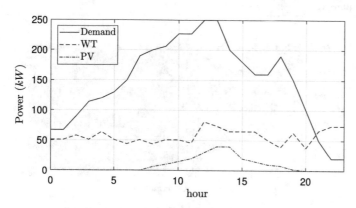

Fig. 6 Power demanded and generated with renewable resources

Table 2 ESS characteristic parameters

SOC^{min} (kWh)	SOC^{max} (kWh)	P_{ESS}^{min} (kW)	P_{ESS}^{max} (kW)	η_c	η_d
70	280	−120	120	0.9	0.9

We consider a known value for the PV and WT generation in the short time of 24 h period. These curves are shown in Fig. 6. The parameters considered for the energy storage system are given in Table 2.

6.2 Genetic Algorithm Settings

The design of the generation schedule of the MG is addressed by means of a GA, using the selection, crossover and mutation rules proposed in Sect. 5. Table 3 contains the main configuration parameters of the GA implementation. It can be noticed that the proposed GA has been tested under different configuration parameters in terms

Table 3 Parameters of the GA

Parameter	Value
λ	3000
μ	3000
Individuals (multi-objective)	3000
Generations	1000
Selection	Tournament size $= 3$
Crossover	Two-point scheme and binary simulated $p_{cx} = [0.6,\ 0.7,\ 0.8]$
Mutation	Gaussian $p_m = [0.4,\ 0.3,\ 0.2]$, $\sigma = 30$
Number of trials	30

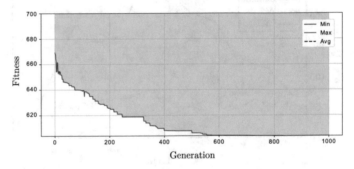

Fig. 7 Evolution of the proposed GA, case-study 1, two-point crossover, $p_c = 0.6$ and $p_m = 0.4$

of crossover and mutation probabilities. In addition, for the SBX method a detailed evaluation of the influence of the eta hyper-parameter has been carried out.

In order to show the convergence of the proposed GA, Fig. 7 depicts the evolution of the individuals throughout the considered number of generations. It is shown that 1000 generations suffices to guarantee convergence. Notice that the convergence of the algorithm can be affected by the value of σ used in the mutation algorithm. Similar convergence degrees have been obtained for the rest of case-studies with the same value of σ.

6.3 Results for the Case Studies

For this problem, different scenarios are proposed, being the results listed below for each of them.

Table 4 Case-study 1 results, expressed in €

SBX							Two-points		
P_{cx}	0.6	0.7	0.8	0.6	0.7	0.8	0.6	0.7	0.8
P_{mut}	0.4	0.3	0.2	0.4	0.3	0.2	0.4	0.3	0.2
eta	0.01			0.1			–		
$best(CF_{ec})$	592.637	595.718	594.888	593.630	594.013	592.637	595.173	597.070	595.863
$mean(CF_{ec})$	602.995	603.793	606.089	603.370	604.439	606.066	603.146	604.071	605.758
$\sigma(CF_{ec})$	5.725	3.769	5.930	4.884	4.065	6.863	4.513	4.553	6.237
P_{cx}	0.6	0.7	0.8	0.6	0.7	0.8			
P_{mut}	0.4	0.3	0.2	0.4	0.3	0.2			
eta	1			10					
$best(CF_{ec})$	594.390	596.204	596.737	596.543	594.180	597.118			
$mean(CF_{ec})$	604.762	604.467	608.589	605.210	604.773	609.207			
$\sigma(CF_{ec})$	6.793	4.449	6.823	6.040	6.059	6.217			

6.3.1 Results for Case-Study 1: Cost-Effective Operation Mode

This case study considers the resolution of the problem by minimizing cost function
(19). The best solutions obtained using the GA for this case-study are listed in Table 4.
Also, in Fig. 8 the best solution for the MG schedule is represented.

Observing Table 4, it can be seen that the proposed approach leads to very similar
solutions, reaching the best one under the SBX scheme with $P_{cx} = 0.6$, $P_{mut} = 0.4$
and $eta = 0.1$. Analyzing Fig. 8, it can be seen that the MT, when is connected to the
MG, generates high amounts of power, which is in accordance with the operation
costs shown in Fig. 4. Analogously, the DE is always providing low power when
connected, reducing in that way the operational costs.

6.3.2 Results for Case-Study 2: Robust Operation Mode

In this second operation mode an $n_r = 3$ level of reliability is guaranteed. That is,
the isolated operation of the MG is guaranteed in the 99.7% of the scenarios. The
objective of the problem is the same than in the cost-effective operation mode, i.e.
minimize the total operation and maintenance cost in the MG.

Table 5 shows the results of the GA for this case study. Note that, in contrast with
case-study 1, the fitness function takes higher values. This fact is a consequence of
the reserve power, that must cover the possible deviation in the demand forecasting.
As can be seen, the best solution is obtained for two-points, when the probabilities
of crossover and mutation are, respectively, 0.8 and 0.2.

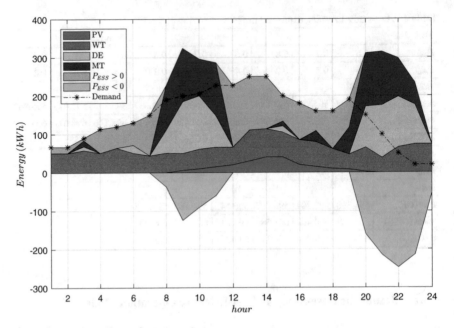

Fig. 8 UC and ED solution for case-study 1

Table 5 Case-study 2 results

SBX						Two-points		
P_{cx}	0.6	0.7	0.8	0.6	0.7	0.8	0.6	0.7	0.8
P_{mut}	0.4	0.3	0.2	0.4	0.3	0.2	0.4	0.3	0.2
eta	0.01			0.1			–		
$best(CF_{ec})$	610.804	613.942	615.287	612.770	611.888	612.615	613.579	614.154	610.792
$mean(CF_{ec})$	619.718	622.898	623.017	621.496	623.111	623.970	622.957	626.400	625.598
$\sigma(CF_{ec})$	4.017	4.198	5.130	4.921	4.604	5.563	4.301	5.578	6.468
P_{cx}	0.6	0.7	0.8	0.6	0.7	0.8			
P_{mut}	0.4	0.3	0.2	0.4	0.3	0.2			
eta	1			10					
$best(CF_{ec})$	613.722	613.636	615.364	613.483	614.783	617.842			
$mean(CF_{ec})$	621.852	625.127	625.648	621.520	624.422	627.119			
$\sigma(CF_{ec})$	4.399	5.210	5.573	5.122	5.773	6.373			

Figure 9 shows the generation schedule for the predicted scenario. That is, considering the possible uncertainties in the demand estimation. It is worth pointing out that, in this case, the battery allows to reduce the amount of energy provided by the MT at every time step increasing in this way the spinning reserve of the MG.

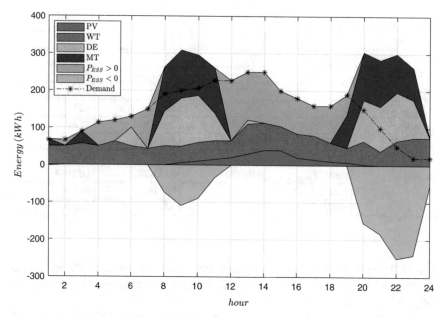

Fig. 9 UC and ED solution for case-study 2

6.3.3   Results for Case-Study 3: Analysis of EES Parameters

In this case-study the effect of the EES parameters in the UC and ED solution are shown. As it was introduced in Sect. 3.7, the EES is characterized by the capacity and the maximum charging and discharging rates. In practice, a battery with both high capacity and fast charging and discharging rates is usually expensive. Then, it is reasonable to achieve a compromise between a good performance in operation and a cost-effective battery.

In this section the resolution of the UC and ED problem is tackled by minimizing the cost function (19) for different levels of reliability as it was done in the previous case-study. However, this time we will consider different models of the EES with the aim of pointing out its influence on the solution of the problem.

Several scenarios are considered in this case:

- Scenario 1: base case. It is considered a $SOC^{min} = 70$ kWh, $SOC^{max} = 280$ kWh and a charging and discharging rate of 120 kW.
- Scenario 2: An increase in the battery capacity is considered with respect to the base case ($SOC^{max} = 350$ kWh).
- Scenario 3: The minimum SOC of the battery is reduced to $SOC^{min} = 20$ kWh with respect to the base case.

Table 6 shows the results of the simulations. 30 simulations have been run for each scenario. It is important to point out that the optimal GA schemes obtained in the previous case-studies have been used. That is, a SBX scheme with $P_{cx} = 0.6$,

Table 6 Case-study 3 results

Scenario		1	2	3
SOC^{min} (kWh)		70	70	20
SOC^{max} (kWh)		280	350	280
P_{ESS}^{min} (kW)		−120	−120	−120
P_{ESS}^{max} (kW)		120	120	120
$n_r = 0$	$best(CF_{ec})$	592.637	593.630	579.64
	$mean(CF_{ec})$	602.995	603.402	586.57
	$\sigma(CF_{ec})$	5.725	4.992	3.885
$n_r = 3$	$best(CF_{ec})$	610.792	615.679	602.988
	$mean(CF_{ec})$	622.957	627.017	611.803
	$\sigma(CF_{ec})$	4.301	5.096	5.670

$P_{mut} = 0.4$ and $eta = 0.1$ when no uncertainty is considered ($n_r = 0$), and a two-points crossover scheme with $P_{cx} = 0.8$ and $P_{mut} = 0.2$ when a robust solution for the problem is pretended ($n_r > 0$).

As it can be seen, for the load and generation profiles considered in the microgrid, an increment in the capacity of the ESS does not have an impact in the performance of the system. However, when the discharging capacity is extended, the solution obtained is clearly better than in the base scenario. Finally, it is important to analyze if the improvements obtained are enough to justify the use of a more expensive ESS.

7 Conclusions

The present study has been carried out for developing a GA to address the problem of finding the most suitable solution for the UC and ED problems in a MG. In sight of the results, several conclusions can be formulated. First, a detailed model of a MG has been introduced to propose a total of 3 case-studies, that have been addressed in order to validate the benefits of the GA to solve the combined problem of UC and ED. When the cost-effective operation mode is considered, the solution prioritizes the use of the DE instead of the MT for low values of demanded power. However, when the MT must be employed, it generates high amounts of power to have an appropriate power/cost ratio. Second, the use of GA allows the user to introduce a more detailed formulation of the problem considering nonlinear terms that approach the model to reality. Third, a new method to approach the UC and ED problems when uncertainties in the problem are taken into consideration has been introduced. This method provides a reliable solution. Note that this fact is crucial when operating a MG in islanding mode. Finally, the effect of the EES parameters in the solution of the problem has been analyzed.

Regarding the limitations of the present work, the computational time required by the GA makes it low efficient to addressing the problem in real-time. In addition, despite the goodness of the obtained results, the absence of guarantee of finding global solutions must always be taken into account, as the algorithm might converge at local minimum. For this reason, an appropriate tuning of the GA parameters is necessary when modifying the problem scenario.

Considering the benefits of the proposed approach, several improvements are proposed as a further work. First, the results of the proposed evolutionary approach will be compared with other metaheuristic algorithms such as Particle Swarm Optimization (PSO) and Evolutionary Strategies (ES), among other techniques. Second, a more sophisticated model of the ESS that considers the effects of battery degradation must be considered. In addition, the implementation of an electricity price short-term forecasting is also proposed to develop a prize-based UC. Finally, the study of this optimization problem in a multi-objective mode is considered as further work. A multi-objective approach would provide a better understanding of the problem.

References

1. Malik, Farhan H., Matti Lehtonen, and Agents in smart grids. 2016. A review. *Electric Power Systems Research* 131: 71–79.
2. Kakran, Sandeep, and Saurabh Chanana. 2018. Smart operations of smart grids integrated with distributed generation: A review. *Renewable and Sustainable Energy Reviews* 81: 524–535.
3. Connor, Peter M., Philip E Baker, Dimitrios Xenias, Nazmiye Balta-Ozkan, Colin J Axon, and Liana Cipcigan. 2014. Policy and regulation for smart grids in the united kingdom. *Renewable and Sustainable Energy Reviews* 40: 269–286.
4. Parhizi, Sina, Hossein Lotfi, Amin Khodaei, and Shay Bahramirad. 2015. State of the art in research on microgrids: A review. *IEEE Access* 3 (1): 890–925.
5. Olivares, Daniel E., Ali Mehrizi-Sani, Amir H Etemadi, Claudio A Cañizares, Reza Iravani, Mehrdad Kazerani, Amir H Hajimiragha, Oriol Gomis-Bellmunt, Maryam Saeedifard, Rodrigo Palma-Behnke, et al. 2014. Trends in microgrid control. *IEEE Transactions on Smart Grid* 5 (4): 1905–1919.
6. Soshinskaya, Mariya, Wina H.J. Crijns-Graus, Josep M. Guerrero, and Juan C. Vasquez. 2014. Microgrids: Experiences, barriers and success factors. *Renewable and Sustainable Energy Reviews* 40: 659–672.
7. Ravichandran, Adhithya, Pawel Malysz, Shahin Sirouspour, and Ali Emadi. 2013. The critical role of microgrids in transition to a smarter grid: A technical review, 1–7.
8. Nuclear Energy Agency. 2019. *The costs of decarbonisation.*
9. Yang, Yuqing, Stephen Bremner, Chris Menictas, and Merlinde Kay. 2018. Battery energy storage system size determination in renewable energy systems: A review. *Renewable and Sustainable Energy Reviews* 91: 109–125.
10. Tan, Xingguo, Qinmin Li, and Hui Wang. 2013. Advances and trends of energy storage technology in microgrid. *International Journal of Electrical Power & Energy Systems* 44 (1): 179–191.
11. Serban, Ioan, and Corneliu Marinescu. 2013. Control strategy of three-phase battery energy storage systems for frequency support in microgrids and with uninterrupted supply of local loads. *IEEE Transactions on Power Electronics* 29 (9): 5010–5020.
12. Wu, Dan, Fen Tang, Tomislav Dragicevic, Juan C. Vasquez, and Josep M. Guerrero. 2014. Autonomous active power control for islanded ac microgrids with photovoltaic generation and energy storage system. *IEEE Transactions on Energy Conversion* 29 (4): 882–892, 2014.

13. Wen, Huiqing, and Weiqiang Zhu. 2016. Control and protection of DC microgrid with battery energy storage system. In *2016 IEEE international conference on power electronics, drives and energy systems (PEDES)*, 1–6. IEEE.
14. Zhanbo, Xu, Xiaohong Guan, Qing-Shan Jia, Wu Jiang, Dai Wang, and Siyun Chen. 2012. Performance analysis and comparison on energy storage devices for smart building energy management. *IEEE Transactions on Smart Grid* 3 (4): 2136–2147.
15. Hatziargyriou, N., H. Asano, R. Iravani, and C. Marnay. 2007. âĂĳ microgrids,âĂİ. *Power and Energy Magazine, IEEE* 5 (4): 78–94.
16. Lasseter, Bob. 2001. Microgrids [distributed power generation]. In *2001 IEEE power engineering society winter meeting. Conference proceedings (Cat. No. 01CH37194)*, vol. 1, 146–149. IEEE.
17. Lasseter, Robert H. 2002. Microgrids. In *2002 IEEE power engineering society winter meeting. Conference Proceedings (Cat. No. 02CH37309)*, vol. 1, 305–308. IEEE.
18. Bullich-Massagué, E., Díaz-González, F., Aragüés-Peñalba, M., Girbau-Llistuella, F., Olivella-Rosell, P., and Sumper, A. 2018. Microgrid clustering architectures. *Applied Energy* 212: 340–361. https://doi.org/10.1016/j.apenergy.2017.12.048.
19. Feng, Wei, Xu Ming Jin, Yi Bao Liu, Chris Marnay, Cheng Yao, and Yu. Jiancheng. 2018. A review of microgrid development in the united states-a decade of progress on policies, demonstrations, controls, and software tools. *Applied Energy* 228: 1656–1668.
20. Castillo, Anya, and Dennice F. Gayme. 2014. Grid-scale energy storage applications in renewable energy integration: A survey. *Energy Conversion and Management* 87: 885–894.
21. Arul, P.G., Vigna K. Ramachandaramurthy, and R.K. Rajkumar. 2015. Control strategies for a hybrid renewable energy system: A review. *Renewable and Sustainable Energy Reviews* 42: 597–608.
22. Nema, Pragya, R.K. Nema, and Saroj Rangnekar. 2009. A current and future state of art development of hybrid energy system using wind and pv-solar: A review. *Renewable and Sustainable Energy Reviews* 13 (8): 2096–2103.
23. Mahlia, T.M.I., T.J. Saktisahdan, A. Jannifar, M.H. Hasan, and H.S.C. Matseelar. 2014. A review of available methods and development on energy storage; technology update. *Renewable and Sustainable Energy Reviews* 33: 532–545.
24. Kazempour, S. Jalal, M. Parsa Moghaddam, M.R. Haghifam, and G.R. Yousefi. 2009. Electric energy storage systems in a market-based economy: Comparison of emerging and traditional technologies. *Renewable Energy* 34 (12): 2630–2639.
25. Khodabakhsh, Raheleh, and Shahin Sirouspour. 2016. Optimal control of energy storage in a microgrid by minimizing conditional value-at-risk. *IEEE Transactions on Sustainable Energy* 7 (3): 1264–1273.
26. Aneke, Mathew, and Meihong Wang. 2016. Energy storage technologies and real life applications-a state of the art review. *Applied Energy* 179: 350–377.
27. Divya, K.C., and Jacob Østergaard. 2009. Battery energy storage technology for power systems an overview. *Electric Power Systems Research* 79 (4): 511–520.
28. Nair, Nirmal-Kumar C., and Niraj Garimella. 2010. Battery energy storage systems: Assessment for small-scale renewable energy integration. *Energy and Buildings* 42 (11): 2124–2130.
29. Poullikkas, Andreas. 2013. A comparative overview of large-scale battery systems for electricity storage. *Renewable and Sustainable Energy Reviews* 27: 778–788.
30. Zia, Muhammad Fahad, Elhoussin Elbouchikhi, and Mohamed Benbouzid. 2018. Microgrids energy management systems: A critical review on methods, solutions, and prospects. *Applied Energy*.
31. Elsayed, W.T., Y.G. Hegazy, F.M. Bendary, and M.S. El-Bages. 2018. Energy management of residential microgrids using random drift particle swarm optimization, 166–171.
32. Alam, Mahamad Nabab, Saikat Chakrabarti, and Arindam Ghosh. 2019. Networked microgrids: State-of-the-art and future perspectives. *IEEE Transactions on Industrial Informatics* 15 (3): 1238–1250.
33. Nikmehr, Nima, and Sajad Najafi Ravadanegh. 2015. Optimal power dispatch of multi-microgrids at future smart distribution grids. *IEEE Transactions on Smart Grid* 6 (4): 1648–1657.

34. Ross, Michael, Chad Abbey, Franois Bouffard, and Gza Jos. 2015. Multiobjective optimization dispatch for microgrids with a high penetration of renewable generation. *IEEE Transactions on Sustainable Energy* 6 (4): 1306–1314.
35. Padhy, N.P. 2004. Unit commitment-a bibliographical survey. *IEEE Transactions on Power Systems* 19 (2): 1196–1205.
36. Nemati, Mohsen, Martin Braun, and Stefan Tenbohlen. 2018. Optimization of unit commitment and economic dispatch in microgrids based on genetic algorithm and mixed integer linear programming. *Applied Energy* 210: 944–963.
37. Shaw, J.J. 1995. A direct method for security-constrained unit commitment. *IEEE Transactions on Power Systems* 10 (3): 1329–1342.
38. Bhardwaj, Amit, Vikram Kumar Kamboj, Vijay Kumar Shukla, Bhupinder Singh, and Preeti Khurana. 2012. Unit commitment in electrical power system-a literature review. In *2012 IEEE international power engineering and optimization conference Melaka, Malaysia*, 275–280. IEEE.
39. Bendotti, Pascale, Pierre Fouilhoux, and Cécile Rottner. 2019. On the complexity of the unit commitment problem. *Annals of Operations Research* 274 (1–2): 119–130.
40. Khan, Aftab Ahmad, Muhammad Naeem, Muhammad Iqbal, Saad Qaisar, and Alagan Anpalagan. 2016. A compendium of optimization objectives, constraints, tools and algorithms for energy management in microgrids. *Renewable and Sustainable Energy Reviews* 58: 1664–1683.
41. Che Ping, and Gang Shi. 2014. An MILP approach for a profit-based unit commitment problem with emissions penalty. In *The 26th Chinese control and decision conference (2014 CCDC)*, 4474–4477. IEEE.
42. Zaree, Niloofar, and Vahid Vahidinasab. An MILP formulation for centralized energy management strategy of microgrids. In *2016 Smart Grids Conference (SGC)*, 1–8. IEEE.
43. Reddy, Srikanth, Lokesh Kumar Panwar, B.K. Panigrahi, Rajesh Kumar, and Ameena Alsumaiti. 2019. Binary grey wolf optimizer models for profit based unit commitment of price-taking genco in electricity market. *Swarm and Evolutionary Computation*, 44: 957–971.
44. Zhang, Di, Songsong Liu, and Lazaros G Papageorgiou. 2014. Fair cost distribution among smart homes with microgrid. *Energy Conversion and Management* 80: 498–508.
45. Wang, Ming Qiang, and H.B. Gooi. 2011. Spinning reserve estimation in microgrids. *IEEE Transactions on Power Systems* 26 (3): 1164–1174.
46. Marquant, Julien F., Ralph Evins, L. Andrew Bollinger, and Jan Carmeliet. 2017. A holarchic approach for multi-scale distributed energy system optimisation. *Applied Energy* 208: 935–953.
47. Fu, Bo, Chenxi Ouyang, Chaoshun Li, Jinwen Wang, and Eid Gul. 2019. An improved mixed integer linear programming approach based on symmetry diminishing for unit commitment of hybrid power system. *Energies* 12 (5).
48. Olivares, Daniel E., Claudio A. Cañizares, and Mehrdad Kazerani. A centralized energy management system for isolated microgrids. *IEEE Transactions on Smart Grid* 5 (4): 1864–1875.
49. Wang, Nan, Lizi Zhang, and Guohui Xie. 2010. An improved mixed integer quadratic programming algorithm for unit commitment. *Dianli Xitong Zidonghua (Automation of Electric Power Systems)* 34 (15): 28–32.
50. Li, Hepeng, Chuanzhi Zang, Peng Zeng, Haibin Yu, and Zhongwen Li. 2015. A genetic algorithm-based hybrid optimization approach for microgrid energy management, 1474–1478.
51. Jo, Kyu-Hyung, and Mun-Kyeom Kim. 2018. Stochastic unit commitment based on multi-scenario tree method considering uncertainty. *Energies* 11 (4): 740.
52. Das, H., B. Naik, and H.S. Behera. 2018. Classification of diabetes mellitus disease (DMD): A data mining (DM) approach. *Computing, Analytics and Networking* 539–549.
53. Sahani, C., R. Rout, J.C. Badajena, A.K. Jena, and H. Das. 2018. Classification of intrusion detection using data mining techniques. *Computing, Analytics and Networking* 753–764.
54. Pradhan, C., H. Das, B. Naik, and N. Dey. 2019. Handbook of research on information security in biomedical signal processing. *Computing, Analytics and Networking* 1–414.
55. Nemati, Mohsen, Karima Bennimar, Stefan Tenbohlen, Liang Tao, Holger Mueller, and Martin Braun. 2015. Optimization of microgrids short term operation based on an enhanced genetic algorithm, 1–6.

56. Askarzadeh, Alireza. 2018. A memory-based genetic algorithm for optimization of power generation in a microgrid. *IEEE Transactions on Sustainable Energy* 9 (3): 1081–1089.
57. Liang, H.Z., and H.B. Gooi. 2010. Unit commitment in microgrids by improved genetic algorithm, 842–847.
58. Li, Peng, Xu Duo, Zeyuan Zhou, Wei-Jen Lee, and Bo Zhao. 2016. Stochastic optimal operation of microgrid based on chaotic binary particle swarm optimization. *IEEE Transactions on Smart Grid* 7 (1): 66–73.
59. Quan, Hao, Dipti Srinivasan, and Abbas Khosravi. 2015. Incorporating wind power forecast uncertainties into stochastic unit commitment using neural network-based prediction intervals. *IEEE Transactions on Neural Networks and Learning Systems* 26 (9): 2123–2135.
60. Ma, Hengrui, Bo Wang, Wenzhong Gao, Dichen Liu, Yong Sun, and Zhijun Liu. 2018. Optimal scheduling of an regional integrated energy system with energy storage systems for service regulation. *Energies* 11 (1): 195.
61. Lasseter, R.H., and P. Paigi. 2004. *Microgrid: A Conceptual Solution* 6: 4285–4290.
62. Hatziargyriou, N., H. Asano, R. Iravani, and C. Marnay. 2007. Microgrids. *IEEE Power and Energy Magazine* 5 (4): 78–94.
63. Fotuhi-Firuzabad, M., and R. Billinton. 1999. Unit commitment health analysis in composite generation and transmission systems considering stand-by units. *IEE Proceedings—Generation, Transmission and Distribution* 146 (2): 164–168.
64. Lei, X., E. Lerch, and C.Y Xie. 2002. Frequency security constrained short-term unit commitment. *Electric Power Systems Research* 60 (3): 193–200.
65. Liu, Guodong, Xu Yan, and Kevin Tomsovic. 2015. Bidding strategy for microgrid in day-ahead market based on hybrid stochastic/robust optimization. *IEEE Transactions on Smart Grid* 7 (1): 227–237.
66. Patel, Jagdish K., and Campbell B. Read. 1996. *Handbook of the normal distribution*, vol. 150. CRC Press.
67. Lasnier, F., and T.G. Ang. 1990. Photovoltaic engineering hand article.
68. Tapia, A., G. Tapia, J.X. Ostolaza, and J.R. Saenz. 2003. Modeling and control of a wind turbine driven doubly fed induction generator. *IEEE Transactions on Energy Conversion* 18 (2): 194–204.
69. Lei, Y., A. Mullane, G. Lightbody, and R. Yacamini. 2006. Modeling of the wind turbine with a doubly fed induction generator for grid integration studies. *IEEE Transactions on Energy Conversion* 21 (1): 257–264.
70. Deshmukh, M.K., and S.S. Deshmukh. 2008. Modeling of hybrid renewable energy systems. *Renewable and Sustainable Energy Reviews* 12 (1): 235–249.
71. Mohamed, F.A., and H.N. Koivo. 2010. System modelling and online optimal management of microgrid using mesh adaptive direct search. *International Journal of Electrical Power & Energy Systems* 32 (5): 398–407.
72. Mohamed, F.A., and H.N. Koivo. 2007. Online management of microgrid with battery storage using multiobjective optimization, 231–236.
73. Ortega-Vazquez, Miguel A., and Daniel S. Kirschen. 2009. Estimating the spinning reserve requirements in systems with significant wind power generation penetration. *IEEE Transactions on Power Systems* 24 (1): 114–124.
74. Ortega-Vazquez, Miguel A., and Daniel S. Kirschen. 2007. Optimizing the spinning reserve requirements using a cost/benefit analysis. *IEEE Transactions on Power Systems* 22 (1): 24–33.
75. Rebours, Yann, and Daniel Kirschen. 2005. What is spinning reserve. *The University of Manchester* 174: 175.
76. Guasch, D., and S. Silvestre. 2003. Dynamic battery model for photovoltaic applications. *Progress in Photovoltaics: Research and Applications* 11 (3): 193–206.
77. Wencong, Su, Jianhui Wang, and Jaehyung Roh. 2013. Stochastic energy scheduling in microgrids with intermittent renewable energy resources. *IEEE Transactions on Smart grid* 5 (4): 1876–1883.
78. Quan, Hao, Dipti Srinivasan, Ashwin M. Khambadkone, and Abbas Khosravi. 2015. A computational framework for uncertainty integration in stochastic unit commitment with intermittent renewable energy sources. *Applied energy* 152: 71–82.

79. Azmy, A., and I. Erlich. 2005. Online optimal management of PEM fuel cells using neural networks, vol. 2, 1337.
80. Orero, S.O., and M.R. Irving. 1997. Large scale unit commitment using a hybrid genetic algorithm. *International Journal of Electrical Power & Energy Systems* 19 (1): 45–55.
81. Wood, A.J., and B.F. Wollenberg. 1984. *Power generation, operation and control.*
82. Arzamendia, Mario, Derlis Gregor, Daniel Gutierrez Reina, and Sergio Luis Toral. 2017. An evolutionary approach to constrained path planning of an autonomous surface vehicle for maximizing the covered area of Ypacarai lake. *Soft Computing* 1–12.
83. Gutiérrez-Reina, Daniel, Vishal Sharma, Ilsun You, and Sergio Toral. 2018. Dissimilarity metric based on local neighboring information and genetic programming for data dissemination in vehicular ad hoc networks (vanets). *Sensors* 18 (7): 2320.
84. John, H. 1984. *Genetic algorithms and adaptation.* Holland, USA: Springer.
85. Luke, Sean. 2009. *Essentials of metaheuristics*, lulu.com, 1st ed. http://cs.gmu.edu/~sean/books/metaheuristics/.
86. Ter-Sarkisov, Aram, and Stephen Marsland. 2011. Convergence properties of two $(\mu + \lambda)$ evolutionary algorithms on onemax and royal roads test functions. arXiv:1108.4080.
87. Deb, Kalyanmoy, and Ram Bhushan Agrawal. 1994. Simulated binary crossover for continuous search space. Technical report.
88. Fortin, Félix-Antoine, François-Michel De Rainville, Marc-André Gardner, Marc Parizeau, and Christian Gagné. 2012. DEAP: Evolutionary algorithms made easy. *Journal of Machine Learning Research* 13: 2171–2175.
89. Gutiérrez, D. 2019. Evolutionary microgrid. https://github.com/Dany503/Evolutionary-Microgrids.

Application of Genetic Algorithms for Designing Micro-Hydro Power Plants in Rural Isolated Areas—A Case Study in San Miguelito, Honduras

A. Tapia, D. G. Reina, A. R. del Nozal and P. Millán

Abstract The use of Micro-Hydro Power Plants (MHPP) has established itself as a fundamental tool to address the problem of energy poverty in rural isolated areas, having become the most used renewable energy source not just in this field but also in big scale power generation. Although the technology used has made important advances in the last few decades, it has been generally applied to big scale hydro-power systems. This fact has relegated the use of isolated MHPPs to the background. In this context, there is still a vast area of improvement in the development of optimization strategies for these projects, which in practice remains limited to the use of thumb rules. It results in a sub-optimal use of the available resources. This work proposes the use of a Genetic Algorithm (GA) to assist the design of MHPP, finding the most suitable location of the different elements of a MHPP to achieve the most efficient use of the resources. For this, a detailed model of the plant is first developed, followed by an optimization problem for the optimal design, which is formulated by considering the real terrain topographic data. The problem is presented in both single (to minimize the cost) and multi-objective (to minimize cost while maximizing the generated power) mode, providing a deep analysis of the potentiality of using GAs for designing MHPP in rural isolated areas. To validate the proposed approach, it is applied to a set of topographic data from a real scenario in Honduras. The achieved results are compared with a baseline integer-variable algorithm and other meta-heuristic algorithms, demonstrating a noticeable improvement in the solution in terms of cost.

A. Tapia · A. R. del Nozal · P. Millán
Departamento de ingeniería, Universidad Loyola Andalucía, Seville, Spain
e-mail: atapia@uloyola.es

A. R. del Nozal
e-mail: arodriguez@uloyola.es

P. Millán
e-mail: pmillan@uloyola.es

D. G. Reina (✉)
Departamento de ingeniería electrónica, University of Seville, Seville, Spain
e-mail: dgutierrez@us.es

© Springer Nature Switzerland AG 2020
M. Rout et al. (eds.), *Nature Inspired Computing for Data Science*,
Studies in Computational Intelligence 871,
https://doi.org/10.1007/978-3-030-33820-6_7

169

Keywords Genetic algorithms · Micro-Hydro Power Plants · Optimization ·
Rural electrification

1 Introduction

The growth of the energy demand around the world constitutes a big challenge
that must be addressed in the coming years [1]. As population and industrialization
grow, global access to energy expands, playing a fundamental role. Nevertheless,
in 2017 there were still 1.06 billion people who lacked access to electricity and 2.6
billion who used biomass to meet their basic needs, in accordance with the World
Data Bank [2]. Although urban areas tend to be more electrified (4% lack access to
electricity), rural areas are the most affected by this problem (27% lack access to
electricity). Furthermore, these statistics are more critical in developing countries,
and thus the expansion of the electricity supply constitutes an area of special interest in
these countries [3], which tends to promote rural electrification programs to improve
life quality of the rural population.

Renewable Energy Sources (RES) play a fundamental role [4] in this context,
having demonstrated to constitute an effective way to guarantee the increasing need of
energy supply (some works predict that by 2050 RES could provide half of the world's
energy needs [5]) without compromising the natural resources, while mitigating
CO_2 emissions. Although a wide range of options have been demonstrated to be
adequate to provide an effective reduction of greenhouse gas emissions, such as
nuclear energy or carbon capture and storage (CCS), the use of RES is praised as one
of the most suitable choices [6]. In contrast with the limitations of nuclear and fossil
fuel availability, RES do not deplete over time, and (with a few exceptions, such as
certain bio-energy production methods and bio-energy life-cycle) are carbon-free.
Nevertheless, the major disadvantages of these systems lie in the uncertainty implied
by the stochastic nature of the natural sources, which translates into a limitation
to high levels of electricity production [7]. Nevertheless, these limitations have no
noticeable effects on small generation systems, making RES ideal candidates to meet
the energy supply requirements for rural areas [8, 9], where due to geographical or
economic reasons, national grid supply are not accessible.

Although there are several alternatives among RES that have been demonstrated
to be suitable to supply electric power to remote isolated areas [10–12], hydro-power
has established itself as the most frequently used [13] since it is capable of reaching
the highest efficiency rates [14] with low investment costs [15]. Given its versatility
and stable projection [16], hydro-power plants represent a suitable and efficient
option to supply rural isolated areas [17].

Nevertheless, despite of the goodness of MHPPs, the precariousness of the context of rural communities usually represents a challenge for the adequate use of
RES. The lack of qualified manpower, together with the limitations of the resources
constitute big barriers to the optimal development of these installations. Within this

context, the study of robust and efficient design strategies is essential to guarantee that the resources are used in the most efficient way, without compromising the limited resources.

1.1 Micro-Hydro Power Plants

Micro-Hydro Power Plants (MHPPs) are hydro-power plants with generation capacities inferior to 100 kW [18]. These systems have small power source requirements and their ease of installation makes them suitable to supply small communities by means of an independent electrical grid [19].

Unlike big-hydro plants, where advance architectures, equipment and expensive civil works are required, MHPPs have both minimal equipment and labour requirements. The water flow is directly extracted from its natural flow without the installation of a reservoir dam. Although a small dam is generally built, its purpose is to guarantee a smooth and clean entrance of the water into the piping. The water is driven downhill through a long pipe (penstock) that ends in the powerhouse, a small building where the generation equipment is installed (a typical scheme is shown in Fig. 1). Inside the powerhouse, the water flow is driven into a turbine, being its kinetic energy

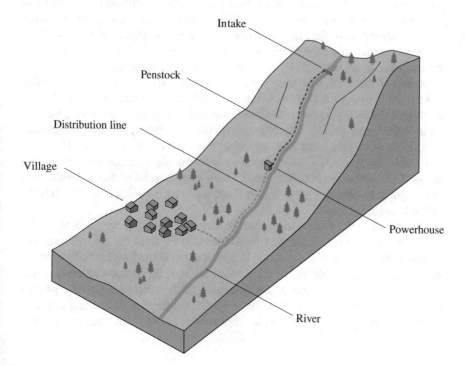

Fig. 1 Basic scheme of a MHPP supplying an isolated village

transformed into mechanical. This last is finally converted into electrical energy by means of a generator, while the water flow is returned to its natural course. Although non-traditional alternatives are frequently used as a turbine, such as locally made systems or pumps working as turbines (PAT) [20], Pelton and Turgo wheels [21] are the usual choices, given their suitability to low flow rates emplacements [22].

1.2 Motivation of the Research

Despite of the cited benefits of MHPP for rural electrification, the precariousness context in which these solutions are developed causes that their true potential is not generally achieved, due to the use of inefficient traditional design methodologies. These methodologies can be generally divided in the following steps:

- Measurement of available flow rate (Q) and height (H_g).
- Estimation of power generation (P).
- Decision-making of dam and powerhouse location.
- Sizing of the equipment.

This traditional process begins with an evaluation of the emplacement, that consists in measuring the available water flow rate and height difference. As flow rate records are not usually available for small rivers and streams, in-situ measurements are generally required (for example, using gauging weirs [23]). Regarding the gross height, its estimation is done by means of traditional methods such as height maps or Topographic Abney levels. With the measurement of flow and gross height, a gross estimation of the obtainable power can be made, in terms of which the feasibility of the plant is evaluated. If the estimations satisfy the requirements considered for the project, the hydro-power professionals proceed to determine the precise location of the water intake (where a small concrete dam is installed) and the powerhouse, on the basis of an initial site visit and know-how. This decision is focused on minimizing the pipe length, L_P, as this variable will strongly condition the cost and the final performance of the MHPP, as will be seen later. In terms of this, the generation system (pipe, turbine and generator) is sized using thumb rules [24], and thus the work labours are planned.

Although it is clear that this routine provides a reasonable approach to design the system, it is clear that the final performance is far from the optimal one. The feasible layout, considering the terrain height, is not evaluated, and the effects on the performance is not considered in the determination of the location of the dam and the powerhouse. For this reason, the estimation of the obtainable power represent is far from the optimal. In sight of these issues, the need of developing practical and efficient design methodologies to improve the use of the natural resources is evidenced.

1.3 Contributions

In order to improve the design of MHPPs to rural electrification projects, this work proposed the application of evolutionary computational approach to address the problem of optimizing the design of a MHPP by finding the most suitable layout, considering the real terrain profile. Although the optimization of MHPP to supply remote areas has been extensively approached in the literature [25–28], most of the studies generally aim at developing general guidelines, while the study of particular design strategies to assist the implementation of MHPP considering the real scenario remains a matter of study. In addition, the complex nonlinear nature of the MHPP performance in a particular location constitutes a high complexity, that has generally been addressed in the literature by simplifying either the domain (approximating the river profile by a straight line [29]), or the problem formulation (fixing certain parameters, such as the pipe diameter [30]). Although it is clear that the existing approaches lead to a better usage of the resources, the cited simplifications imply that the obtained solutions may differ from the optimal solutions of the real problem. In this work, a Genetic Algorithm (GA) is developed to find the most suitable layout of a MHPP, on the basis of the topographic characteristics of the emplacement. To this end, the terrain profile is obtained through a topographic survey, in terms of which the problem is formulated. The paper aims to find the optimal location of the main elements, this is, the powerhouse, the dam and the distribution of pipe lengths along the terrain, including the selection of the most adequate penstock diameter, in accordance to a set of performance criteria. In addition, to obtain a deeper understanding of the potential of the emplacement, the multi-objective problem is also studied, being three competitive objectives of the model optimized simultaneously. With this, the influence of the different parameters in the performance of the MHPP is studied. A real river profile scenario in Honduras is applied to verify the benefits of the proposed approach.

This chapter is organized as follows: A general overview of the related work in the literature is presented in Sect. 2, where the main advantages of the proposed within the actual state of the art approach are presented. In Sect. 3, a detailed description of the problem set-up and the main variables is made. For this, a model of the MHPP is developed in terms of the decision variables, being the model used then to define the problem of optimally designing the MHPP layout. In Sect. 4, a GA is developed to address the optimization problem proposed. In Sect. 5, the proposed GA is applied to a real scenario in Honduras, where the results are summarized and the performance of the approach is validated. Finally, the conclusions of this work, together with the concluding remarks, are summarized in Sect. 7.

2 Related Work

During the last years, the optimization of MHPP has received a big attention in the literature [31–34]. A deep analysis of the current state of the art in computational optimization methods applied to RES can be found in [34]. In this relevant work, the latest research advances, regarding he different optimization strategies applied to the design of RES systems, are summarized. A similar more recent work is [33], where the authors review the main optimization methods for the deployment and operation of RES generation units. In general, the complexity of reliably modeling RES systems, especially hydro-power plants [26, 35], has motivated the decomposition of the problem on the basis of specific aspects, such as determining the most suitable operation strategies [36, 37] or sizing the equipment [28]. The main reason lies in the use of analytical approaches, whose performance is strongly conditioned by the complexity of the problems addressed. For this reason, it is usual to simplify the problems formulation as much as possible. For example, the authors in [29]study the optimization of penstocks in MHPPs in order to minimize the energy cost, simplifying the river by means of an average slope. This same approximation is considered in [38], where the optimal flow discharge and penstock diameter are determined by means of a dimensional analysis. In this approach the problem aims to minimize the water usage, being also a set of dimensionless relationships between the relevant design variables derived. It is important to note that the flow rate respond to a stochastic nature, and thus a stochastic approach can be proposed to improve the design of MHPPs. For example, the authors in [39] consider the Flow Duration Curves (FDC) and the environmental requirements to develop an analytical framework to determine the performance and profitability of a MHPP, being it validated in a real case. Although the use of FDC has been relevant in the literature, its application to small scale plants is not generally relevant, being its potential generally focused in bigger scales. An example of this is [40], where the authors present a toolbox to optimize the design of hydro-power plants by means of performance simulations in terms of FDC.

Although traditional optimization approaches such as Linear Programming (LP) [41], Integer Linear Programming (ILP) [30] or Mixed-Integer Non-Linear Programming (MINLP) [41] have been proven useful to address these problems, meta-heuristic algorithms are gaining relevance in several areas of engineering [42, 43], and especially in the field of systems optimization [44–46]. An illustrative example of this can be found in [45], where the authors develop a numerical sizing tool on the basis of simulations of the plant performance (production and cost) during the year. In terms of these simulations, a parametric study is developed to evaluate the effects of the different factors, by means of a stochastic evolutionary algorithm implemented. In [46], the authors propose three different to optimize reservoir operation and water supply operations. Similarly, a strategy for the optimal design, control and operation of MHPPs is proposed in [27], where the authors present a Honey Bee Mating Optimization (HBMO) algorithm. This algorithm determines the turbine type and number, and the penstock diameter, in addition to scheduling the operation that maximizes the benefit for a given set of FDC. Nevertheless, this approach does not consider particular characteristics of the plant location. Similar to the present work but taking into account this last issue, the MHPP layout is optimized in [47], where the authors develop a Genetic Algorithm (GA) to find the most adequate locations for the different parts of the plant, including powerhouse, dam and penstock layout, in order to reach a certain power rate with the minimal cost, satisfying a set of constraints related to flow usage and feasibility of the layout for a certain terrain profile.

This work proposes the use of a GA [48] to find the optimal layout of MHPP in a certain location, which is considered as an input by means of a topographic survey. The problem is formulated on the basis of the framework proposed in [30]. An improved cost function is proposed, and the generated power, flow usage and physical feasibility constraints are considered. To verify the benefits of the proposed approach, it is applied to optimize a real MHPP project in a rural community of Honduras.

3 Problem Statement

The layout of a MHPP constitutes a strong conditioning to its performance, and thus finding its optimal configuration represents a challenge itself, as requires a compromise between different parameters, such as the gross height and the length of the penstock. The higher the gross head is, the bigger the amount of obtainable power is. Nevertheless, a long penstock implies negative effects due to the friction losses, especially for small penstock diameters [49]. For this, a model of the plant is required to be developed.

3.1 Model of the System

The objective of modeling the MHPP is based on finding the relation between the variables that result from the plant layout (this is the gross head H_g and the penstock length L_P) and the variables that determine the performance of the plant (this is the generated power P, the water flow rate Q and the cost C).

3.1.1 Generated Power

The power obtained in a MHPP can be expressed as

$$P = Qh\eta, \tag{1}$$

Fig. 2 Working scheme of a typical Pelton microturbine

being η the efficiency of the generation equipment (turbine and generator) and h the water height at the entrance of the turbine. The difference between the gross height, H_g, and this last relies in the heigh losses, h_L, that appear due to the friction. This is

$$h = H_g - h_L. \tag{2}$$

As the turbines considered are action micro-turbines [26], the energy conversion is made by means of an atmospheric jet (see Fig. 2), and thus the energy of the flow at the entrance of the turbine is entirely kinetic. Following this, using v_{jet} to denote the water speed of the jet, it can be written that

$$h = \frac{1}{2g}v_{jet}^2. \tag{3}$$

Given the incompressibility of water, the speed of the jet v_{jet}, can be expressed in terms of the flow Q, and the sectional area of the nozzle injector S_{noz} as

$$v_{jet} = \frac{Q}{c_D S_{noz}}, \tag{4}$$

where the coefficient of discharge c_D models the formation of a jet contraction after the water leaves the nozzle [24]. Using this last expression, the height at the entrance of the turbine h can be written in terms of the flow as

$$h = \frac{1}{2g c_D^2 S_{noz}^2} Q^2. \tag{5}$$

With respect to the friction loss h_L, it can be modeled by using several approaches proposed in the literature. In this work, the same expression used in [30] is used, yielding that

$$h_L \approx k_{fric} \frac{L_P}{D_P^5} Q^2, \tag{6}$$

where k_{fric} is a constant that depends on the material, D_P is the pipe diameter. Introducing expression (4) in (5), and the resultant expression, together with (6), in (3), the following expression for the flow Q can be obtained:

$$Q = \left[\frac{H_g}{\frac{1}{2gc_D^2 S_{noz}^2} + \frac{k_{fric}}{D_P^5} L} \right]^{\frac{1}{2}}. \tag{7}$$

Finally, an expression for the obtainable power P, in terms of the height H_g and length of the penstock L_P can be obtained by substituting (7) and (5) in (1), resulting

$$P = \frac{\eta\rho}{2S_{noz}^2} \left[\frac{H_g}{\frac{1}{2gS_{noz}^2} + k_{fric} \frac{L}{D_P^5}} \right]^{\frac{3}{2}}. \tag{8}$$

3.1.2 Cost of the Plant

Given the context of precariousness presented in Sect. 1, the cost of the installation constitutes the main limitation for the success of the project of developing a MHPP for rural electrification. It is relevant to note that the overall cost of the installation includes not just the penstock, the dam, and the powerhouse, but also the distribution line, the turbine, and the generator. Despite of this, the sizing of the turbine and generator is usually conditioned by the order of the power estimation at each location. The device selection is made for a wide margin of operation points, and thus its dependency on the nominal point does not present a significant difference in the manufacturing process, as the cost is approximately constant. For this, the location of the dam and the powerhouse, together with the layout of the penstock and the power line represent the main conditioning factors regarding the problem of optimizing the use of the funding resources. As the cost of the generating equipment is considered constant in this work, it is then not considered in the optimization problem.

The cost function considered is defined as the sum of the cost of the penstock, C_P and the cost of the distribution line C_E. This is

$$C = C_P + C_E. \tag{9}$$

The price of a piping installation typically varies in proportion to the total weight of the pipes, which depends not only on the length but also on the diameter D_P and the wall width. Following [29], a simple expression for the cost of the pipe has the form of

$$C_P = k_P L_{P,eq} D_P^2,$$

being K_P a constant that must be defined in terms of the pipe material and its geometry, and $L_{P,eq}$ is the equivalent length of the penstock, which is defined as the physical length plus a virtual length to consider the drawbacks of installing and deploying the connections between the pipe lengths. Using λ to model the equivalent cost of installing each elbow, it can be written

$$L_{P,eq} = L_P + \lambda n_c,$$

and thus the cost of the penstock is

$$C_P = k_P(L_P + \lambda n_c)D_P^2. \tag{10}$$

With respect to the cost of the distribution line, it can be modeled as

$$C_E = k_E L_E, \tag{11}$$

being k_E the linear cost of the wiring. Introducing (10) and (11) in (12), the cost function is

$$C = k_P(L_P + \lambda n_c)D_P^2 + k_E L_E. \tag{12}$$

The distribution line of the MHPP is assumed to consist of two intervals. The first interval connects the village with the closest point of the river, named s_c, while the second connects this point to the powerhouse along the river. This scheme is followed by the difficulties of access that the rough terrain imply, easing its installation and maintenance. As the first interval is not conditioned by the layout of the plant, only the second one is considered in the optimization problem. Note that the location of this point s_c must be determined in advance.

3.2 Model of the Layout

The possible feasible layouts are modeled by using the method proposed in [30], where the domain of the problem is defined as a N−discretization of the topographic profile of the river, in the form of

$$\{s_i, z_i\}, \qquad i = 1 \dots N, \tag{13}$$

Variables s_i and z_i are the coordinates of the i-th point belonging to the profile of the river, represented in the plane $s - z$ that results from the 2D-development of the real profile. This 2D development (schematically represented in Fig. 3) is made on the basis of several assumptions. First, given the mountainous geography related to the remote nature of the studied areas, the water sources are upper-course rivers, typically formed in V-shaped valleys with very low or negligible curvature. This, in addition with the short length of the required penstocks, implies that neglecting the 3D nature of the layout will have no noticeable implications on the formulation of the problem.

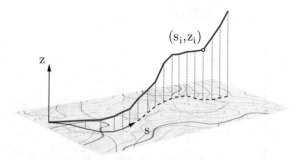

Fig. 3 Scheme of the 2D simplification of the 3D real river profile

(s_i, z_i)

The solutions are modeled by means of a variable Δ is defined as a set of N binary variables δ_i in the form of

$$\Delta = [\,\delta_i\,]_{1 \times N} \qquad i = 1 \ldots N, \tag{14}$$

These variables δ_i are defined in such a way that each combination of them (resulting in an array Δ) defines a layout of the MHPP as follows:

- A value $\delta_i = 1$ represents the deployment of an elbow in point (s_i, z_i).
- The minimal index i satisfying $\delta_i = 1$ represents the location of the powerhouse.
- The maximum index i satisfying $\delta_i = 1$ represents the location of the dam.

For a better understanding of the proposed scheme, an illustrative example of this scheme is represented in Fig. 4 for a discretization with $N = 10$ and an arbitrary Δ. Note that the number of combinations for Δ is 2^N. With this in mind, it must be noted that even a low sampled set of data with a $N \sim 100$ (this corresponds to a 1 km river with a height measurement each 10m) represents an intractable problem for brute-force algorithms, justifying the use of meta-heuristic approaches.

Using this scheme, the variables H_g, L_P and L_E can now be defined in terms of the solution Δ. The gross head is defined as the height difference of the locations of the dam and the powerhouse, and thus it can be determined as the height difference between the maximum and minimum points (s_i, z_i) that satisfy $i = 1$. This is

$$H_g = z_d - z_p, \tag{15}$$

where the sub-index d and p refers, respectively, to the maximum and minimum index i such that $\delta_i = 1$. Thus these index corresponds to the water dam and the powerhouse location, respectively.

Regarding the length of the penstock, L_P, it can be determined as the summation of the successive pipe lengths of between consecutive elbows, this is

$$L_P = \sum_{\forall (i,j)} \left[(s_j - s_i)^2 + (z_j - z_i)^2 \right]^{\frac{1}{2}} \qquad \forall\, (i,j) \quad \left| \begin{array}{l} \delta_i = \delta_j = 1, \\ \delta_k = 0 \ \forall k \in \{i, j\}. \end{array} \right. \tag{16}$$

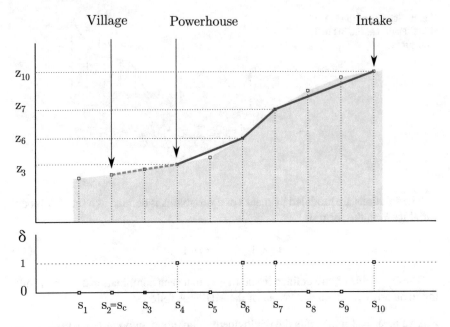

Fig. 4 Illustrative example of the layout modeling, considering an arbitrary terrain (blue dashed line) discretized by means of $N = 10$ points (squares). The arbitrary solution $\Delta = [0\,0\,0\,1\,0\,1\,1\,0\,0\,1]$ has been represented (penstock in red continuous line, distribution line in orange dashed line). Note that the connection point to the village is located in $i_c = 2$

A similar analysis can be made to determine the length of the distribution line, L_E, which can be calculated as the length of the river interval between the powerhouse location and the nearest point to the village, s_c. This is

$$L_E = \sum_{i=min(i_c,i_p)}^{max(i_c,i_p)} \left[(s_{i+1} - s_i)^2 + (z_{i+1} - z_i)^2 \right]^{\frac{1}{2}}. \tag{17}$$

Please note that this expression covers the possibilities of the village connection point, s_c, being lower than the powerhouse and the opposite.

3.3 Formulation of the Problem

In this work, the problem consists of finding the best location of the powerhouse, and the water intake, together with the diameter and layout of the penstock. This problem is formulated in single (SO) and multi-objective (MO) modes. The SO problem is formulated as a cost minimization problem, while the MO additionally pursuits the maximization of the generated power.

The constraints applied to the problem cover the minimal energy supply required for the village, the maximum amount of water which is allowed to be extracted, and the feasibility of the penstock deployment on the terrain. These constraints are detailed below.

3.3.1 Power Constraint

The fundamental requirement of the MHPP relies in the need of a certain power level that satisfied the considered energy needs of the village. To this end, a value of the necessary power supply, P_{min}, is established on the basis of the initial evaluation of the community supplied, and usually covers illumination and shared household appliances. The constraint is then established as

$$P \geq P_{min}$$

This expression can be transformed by introducing Eqs. (15) and (16), resulting in

$$\frac{\eta \rho}{2 C_D^2 S_{noz}^2} \left[\frac{z_d - z_p}{\frac{1}{2 g c_D^2 S_{noz}^2} + \frac{k_{fric}}{D_p^5} \sum_{\forall (i,j)} \sqrt{(s_j - s_i)^2 + (z_j - z_i)^2}} \right]^{\frac{3}{2}} \geq P_{min}. \qquad (18)$$

3.3.2 Flow Constraint

The water flow available in the stream constitutes a strong limitation to the size of the MHPP, and thus the turbine flow Q must be limited by an maximum extraction, which is defined as a fraction κ of the natural flow rate Q_{river}. This constraint is introduced in the form of

$$Q \leq \kappa Q_{river},$$

This equation can be formulated in terms of the variables of the decision variables by introducing (15) and (16), resulting in

$$\left[\frac{z_d - z_p}{\frac{1}{2 g c_D^2 S_{noz}^2} + \frac{k_{fric}}{D_p^5} \sum_{\forall (i,j)} \sqrt{(s_j - s_i)^2 + (z_j - z_i)^2}} \right]^{\frac{1}{2}} \leq \kappa Q_{river}. \qquad (19)$$

3.3.3 Feasibility Constraints

In order to guarantee that the obtained solutions are feasible, a set of constraints are proposed in such way that the height differences between the penstock and the terrain are small enough to be covered by using supports (see point 5 in Fig. 4) or excavations (see points 8 and 9 in Fig. 4), respectively. This is done by means of two different constraints:

- The pipe can be disposed at a certain height from the terrain profile only if this distance is smaller than the maximum available length of the supports, denoted by ϵ_{sup}.
- The pipe can be disposed under a certain depth from the terrain profile only if this distance is smaller than the maximum depth of excavations that is possible to be made, denoted by ϵ_{exc}.

Note that the both ϵ_{exc} and ϵ_{sup} must be estimated on the basis of the properties of the terrain. Using $z_{P,i}$ to denote the height of the penstock at coordinate s_i, the feasibility constraints are written as

$$z_{P,i} - z_i \leq \epsilon_{sup} \quad \forall i = 1 \ldots N, \tag{20a}$$

$$z_j - z_{P,j} \leq \epsilon_{exc} \quad \forall j = 1 \ldots N. \tag{20b}$$

4 Evolutionary Computational Approach

Evolutionary algorithms are meta-heuristic approaches capable of achieving significant results in complex optimization problems [48]. Although many evolutionary algorithms have been proposed in the literature [50], Genetic Algorithms (GAs) have been especially relevant in many engineering optimization problems [51–53]. GAs are population and nature-inspired approaches that are based on the process of natural selection. The main idea of GAs lies in the encoding of the solutions in a chromosome-like structure, in which each optimization variable is codified as a gen.

The working principle of a GA consists of creating a set of different individuals (population), that represents different potential solutions of the problem. These individuals evolve through a certain number of iterations named generations (see Fig. 5), successively creating new offspring that adapts better to the optimization landscape, in accordance with the survival principle of the Darwinian theory. The adaptation of an individual is measured through its quality or fitness, this is, the evaluation of the solution as an input of the objective function of the problem.

A set of genetic operators, such as selection, crossover and mutation, are used to create the offspring. The selection is an elitist operation that is based on selecting the parents that participate in the crossover and mutation operations. Therefore, individuals with higher quality have more probability of being selected as parents. The

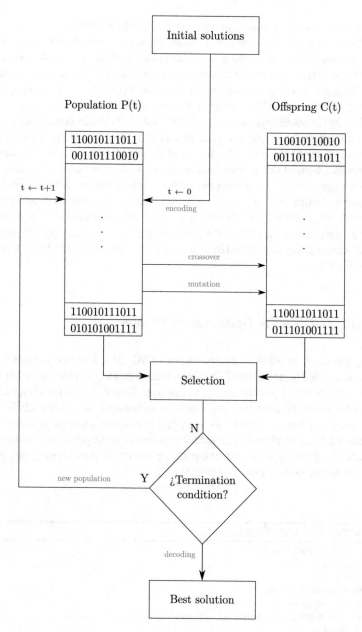

Fig. 5 Working principle of a genetic algorithm

crossover operation is based on combining the genetic information of two different individuals, creating other two new individuals. With respect to the mutation operator, it consists in changing the genetic information of an individual to generate a new one. Both crossover and mutation are probabilistic operations, and are capable of developing a good exploration and exploitation capabilities with a proper tuning.

The application of a GA to address the proposed optimization problem is based on codifying the possible layouts of the MHPP into individuals, containing the information related to the location of powerhouse and dam, the layout of the penstock and the diameter of the pipe. The fitness of the individuals is defined through the objective functions of the optimization problem. As was previously mentioned, two different configurations have been considered such as SO and MO approaches. In the single-objective case, the fitness function is the cost of the plant, and thus it is a minimization problem. On the contrary, in the MO case the GA simultaneously minimize the cost and maximizes the power. In this second case, the multi-objective case, the optimization is achieved by applying the Pareto dominance-based technique NSGA-II [54].

4.1 Single-Objective Optimization Problem

Among the many possible implementations of SO GAs, a mupluslambda scheme [55] (summarized in Algorithm 1) has been used in this work. This algorithm begins with a random initial population P_i, which is evaluated. Then, the offspring μ is created by means of crossover and mutation operations, whose probabilities are, respectively, p_{cx} and p_{mut}. Next, the offspring is evaluated, and the new population λ is selected from the offspring generated together with the previous population P_g. The benefits of this approach lie in the strong elitism, as the offspring and parents have to compete each other to be selected [55].

Algorithm 1: GA mupluslambda.

---
1 Create initial population P_i;
2 Evaluate P_i;
3 $P_g = P_i$;
4 **while** $stop == False$ **do**
5   Parents' selection;
6   Create offspring μ (crossover p_{cx} and mutation p_{mut});
7   Evaluate μ ;
8   Select new population λ $(\mu + P_g)$;
9     $P_g = \lambda$;
10 **end**

The stop criteria of the GAs is generally based on a certain number of generations, after which the algorithm finishes. The final population contains the best solutions for the optimization problem.

4.1.1 Individual Representation

Each individual representing a possible design of the MHPP is a list containing the ones or zeros (binary variables) that constitute the solution Δ (according to (14)). Also, the diameter D_P of the penstock is also embedded in the chromosome in the form of a 5 bits codification, assuming that the penstock has the same diameter along its layout. The length S of the chromosome is either N or $N + 5$ depending on whether the diameter of the pipes is considered or not (see Sect. 5 for more details). The first N bits correspond to the discrete data obtained from the profile of the river (see the scheme in Fig. 6a). When the diameter D_P is considered as an optimization variable, it is embedded in the chromosome by the last five bits (see Fig. 6a). Using this codification for D_P, 32 decimal numbers can be represented. As the diameter cannot be equal to 0, the decimal numbers represented are within the interval $\{1 - 32\}$, determining the value in centimeters of the diameter D_p.

Although random generation is used to create the initial population of the GA, in order to provide feasible solutions in the initial step and avoiding discarding invalid ones, a tailored individual generator is proposed. This algorithm consists of selecting two random points (p_1 and p_2) within the interval $[1, S]$ (with $p_1 < p_2$), and filling up with ones the variables δ_i within the interval $[p_1, p_2]$ (see Fig. 6).

With this approach, the individuals of the initial population will not suffer from the demanding feasibility constraints described in Sect. 3.3, favouring an efficient exploration during the first generations of the GA.

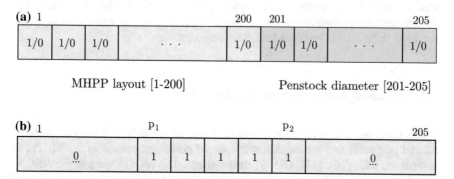

Fig. 6 Individual representation considering the diameter of the penstock D_p (**a**) and individual generation scheme (**b**)

4.1.2 Fitness Function

The cost function defined in expression (12) determines the fitness of each individual, in such way that the lower the cost, the better the solution is. In order to discard the invalid solutions and avoid their participation in the following generations, death-penalty is used. Following this, the fitness of each individual will be calculated as

$$\begin{cases} if \text{ \textbf{solution valid}} \ F = (8), \\ \quad\quad else \quad\quad\quad F = -\infty, \infty, \quad according \ to \ (18), (19), (20). \end{cases}$$

4.1.3 Genetic Operators

Given the good results that the tournament selection mechanism has demonstrated in the literature [56], it has been used in this work. This mechanism consists in selecting a number of individuals that compete each other to be chosen as a parent, being the best one selected as a parent to participate in the crossover and mutation operations [56]. The size of the tournament has been defined as three, as it has been proven suitable for a wide variety of problems.

Regarding the crossover operation, a two-points scheme is used. This consists of swapping the genetic information of the parents by mean of two points that act as indexes of the exchange. With respect to the mutation algorithm, a tailored method has been proposed. This consists in a modified flip-bit method, in which the probability of flipping a one to a zero is considerable higher than a zero to a one. The aim of this method is to reduce the cost of the layout by reducing the number of elbows of the penstock and thus its length. It is relevant to note that, given the proposed scheme for the initial generation, these individuals will have a high number of ones, and thus, a high number of elbows which translates into a high cost. With this mutation scheme, the cost is expected to be gradually reduced through the generations. For this scheme, the probability of converting a one to a zero, p_{hl}, and a zero to a one, p_{lh}, have to be established (see Sect. 5 for more details).

4.2 Multi-objective Optimization Problem

In this case, the objective is simultaneously the minimization of the cost (electrical and penstock) and the maximization of the power generation of the MHPP, in accordance with (10), (11), and (8), respectively. The MOGA used is the NSGA-II [54], based on the Pareto dominance and that has demonstrated to be capable of providing good results for a wide range of optimization problems in engineering [57]. The Pareto dominance establishes that a solution dominates another iff it is strictly superior in all considered objectives. Therefore, the aim of the NSGA-II is to find all non-dominated solutions, which form the so-called Pareto front. For the optimization problem addressed in this work, the Pareto front will consist of a curve in the 2D plane cost-power. The implementation of the MO-GA is summarized in Algorithm 2.

The main differences with respect to the single-objective case are the following:

i The evaluation of the individuals for the two objectives is required to be calculated
ii The selection mechanism is based on the Pareto dominance. For this reason, the Pareto front is updated after every generation, being this composed by all the non-dominated solution when the algorithm finishes.

Algorithm 2: GA based on NSGA-II.

---

1 Create initial population P_i;
2 Evaluate P_i;
3 $P_g = P_i$;
4 **while** $stop == False$ **do**
5   Parents' selection;
6   Create offspring μ (crossover p_{cx} and mutation p_{mut});
7   Evaluate μ;
8   Calculate dominance;
9   Update Pareto front;
10   Select new population based on dominance λ ($\mu + P_g$);
11      $P_g = \lambda$;
12 **end**

Using the MO approach provides the decision maker with a big picture of the potential of the emplacement, allowing the consideration of different layouts if required (changes in the budget, power requirements, etc).

4.2.1 Individual Representation

The individual representation is the same as that considered in the SO case.

4.2.2 Fitness Function

In this case, the fitness of each individual is a tuple of three components (one for each objective). As was done in the SO mode, the death penalty is applied in order to penalize the invalid solutions. Note that the death penalty must be employed in each of the considered objectives, this is

$$\begin{cases} \text{if } \textbf{solution valid } F = (8), (10), (11), \\ \qquad \text{else} \qquad F = \infty, \quad \text{according to } (18), (19), (20). \end{cases} \qquad (21)$$

188 A. Tapia et al.

4.2.3 Genetic Operators

The crossover and mutation schemes used are the same of the single-objective case.

5 Simulation Results

In this section, the proposed evolutionary approach is applied to a real location. The selected scenario consists of designing a MHPP to supply the small village of San Miguelito, in the Department of Santa Bárbara (Honduras). This rural remote community gathers ideal environmental characteristics, and given its location it lacks access to the national electrification grid. A topographic survey of the terrain is used in this Section. Also, a satellite image of the emplacement has been obtained, being i shown in Fig. 7.

5.1 Scenario Settings

The topographic survey provided a discretization of $N = 67$ points (represented in Fig. 8 for a better understanding), in the form of (13).

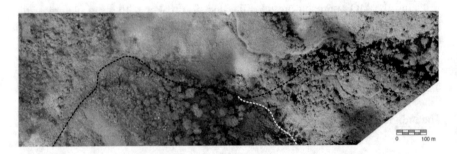

Fig. 7 Aerial view of the studied river profile (black) and a small tributary (white)

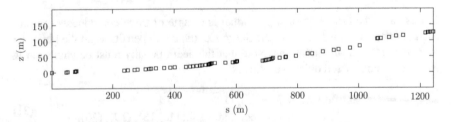

Fig. 8 Topographic data points (squares)

Table 1 Parameters of the case study

Parameter	Symbol	Value	Units
Minimal power requirement	P_{min}	8	kW
Flow of the natural course	Q_{river}	50	L/s
Maximum flow extraction allowance	κ	0.5	–
Equivalent cost of elbows	λ_c	50	m
Maximum depth of excavations	ϵ_{exc}	1.5	m
Maximum height of supports	ϵ_{sup}	1.5	m
Cost coefficient of the pipe	k_P	700	$/m^3$
Cost coefficient of the distribution line	k_E	22	$/m

The objective of this problem is designing a MHPP to supply the community with a basic power need of 8 kW. The flow of the river is 50 l/s, and a 50% fraction is considered allowed to be extracted. The characteristics of the terrain permits the installation of supports and excavations up to 1.5 m. In addition, an equivalent cost of $\lambda_c = 50$ m is considered for the pipe elbows. The selected commercially available pipe is 600 kPa uPVC, which is suitable for pressures up to 87 m [29]. For this pipe, the coefficient has been estimated as $K_P = \$700/m^3$. With respect to the distribution line, a typical value of $k_E = \$20/m$ has been used. Note that these parameters consider the additional related costs (transport, deployment, etc). These parameters have been summarized in Table 1.

5.2 Genetic Algorithm Settings

The problem proposed is addressed by means of a GA, which employs the crossover and mutation operators proposed in Sect. 4.[1] Table 2 contains the main configuration parameters of the GA implementations.

5.3 Results

5.3.1 Single-Objective Mode

This case study represents the basic problem of optimizing a MHPP, regarding the location of the powerhouse and the dam, the layout of the penstock and its elbows, and the diameter of the pipe used. To determine the most suitable parameters of the GA, the influence of the crossover and mutation probabilities, this is p_{cx} and p_{mut},

---

[1]The code is available in [58]. The simulator has been developed using Python and DEAP [59].

Table 2 Parameters of the GA

Parameter	Value
λ	2000
μ	2000
Individuals (MO)	2000
Generations	100
Selection	Tournament size $= 3$ (SO) NSGA-II (MO)
Crossover	Two-point scheme $p_{cx} = [0.6, 0.7, 0.8]$
Mutation	Modified bit-flip $p_m = [0.4, 0.3.0.2]$, $p_{hl} = 0.8$, $p_{lh} = 0.2$
Number of trials	30

Table 3 Results of the SO problem

GA parameters				
p_{cx}	0.5	0.6	0.7	0.8
p_{mut}	0.5	0.4	0.3	0.2
Final population				
Mean fitness	10572	11403	12714	13593
Std. dev. fitness	1665.4	2243.6	2800.2	2788.1
Best individuals				
Gross height (m)	94.76	94.76	94.76	94.76
Flow rate (L/s)	13.716	13.716	13.716	13.716
Power (kW)	8.035	8.035	8.035	8.035
Penstock length (m)	753.15	753.15	753.15	753.15
Distribution line length (m)	21.39	21.39	21.39	21.39
Number of nodes	12	12	12	12
Pipe diameter (cm)	10	10	10	10
Penstock cost ($)	9471.9	9471.9	9471.9	9471.9
Distribution line cost ($)	470.58	470.58	470.58	470.58
Total cost (Fitness) ($)	9942.5	9942.5	9942.5	9942.5

has been evaluated by using the values listed in Table 2. The results obtained are summarized in Table 3. In addition, the layout corresponding to the best individual has been represented in Fig. 9.

In sight of the results listed in Table 3, some comments can be made. First, it can be seen that the four combinations of crossover and mutation probabilities led to the same optimal solution. With respect to the optimal solution, the dominance of the distribution line over the cost of the penstock in the overall cost is evidenced, as the powerhouse is located next to the connection point. This can be understood by means of the versatility of the penstock layout with respect to the distribution line. Displacing the penstock layout along the domain result in small fluctuations of the penstock

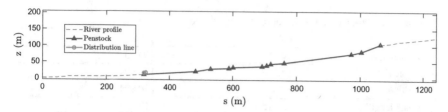

Fig. 9 Best solution obtained for SO problem. The penstock (black line), the elbows (black triangles) and the distribution line (orange line) are represented. In addition, the raw data points are also plotted (blue circles), together with a linear interpolation of these (blue dashed line)

cost (indeed two different penstocks that cover different areas of the terrain have the same cost). This does not happens with the distribution grid, as its costs is strongly conditioned by the location of the powerhouse. Nevertheless, the combination of $p_{cx} = 0.5$, $p_{mut} = 0.5$ has demonstrated a better overall performance, with a mean fitness of \$10572.

5.3.2 Comparison with Other Algorithm

In this section, the optimization problem proposed is addressed by means of the ILP proposed in [30], where the authors develop a linear formulation of the problem, to be solver by a Branch and Bound Algorithm (BBA). Although the linear formulation permits noticeably shorter solving times, the non-linear nature of the distribution grid definition (see expression 17) and the effects of the penstock (see expression 6) cannot be modeled, and thus, these are not able to be considered in the ILP problem. Nevertheless, the optimal diameter, $D_p = 10$ cm, obtained by the GA is proposed for the ILP problem. The optimal layout obtained by this approach is represented in Fig. 10. Observing this last, it can be seen as the consideration of the cost of the distribution line, C_e, can constitute a conditioning for the problem. For the sake of comparison, the main variables of this solution are compared with the best solution obtained by the GA in Table 4.

It can be seen that the ILP approach provides a better (cheaper) penstock. This is due to the non consideration of the distribution grid, which strongly conditions

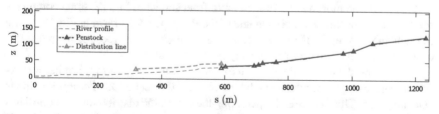

Fig. 10 Best solution obtained for SO problem obtained by using BBA [30]

Table 4 Comparison of the results of the GA and the BBA

	GA	BBA
Gross height (m)	94.76	94.45
Flow rate (l/s)	13.716	13.974
Power (kW)	8.035	10.994
Penstock length (m)	753.15	653.01
Distribution line length (m)	21.39	282.34
Number of nodes	12	10
Pipe diameter (cm)	10	10
Penstock cost ($)	9471.9	8071
Distribution line cost($)	598.96	6211.5
Total cost ($)	9942.5	14282

the location of the powerhouse. As was expected, in spite of the more economic penstock, the solution obtained by the ILP provides a much more expensive layout, as the cost of the distribution grid is noticeably higher than in the GA solution.

In conclusion, the results demonstrate the capability of the proposed GA to address the complex problem of finding the optimal layout of the MHPP with good results.

5.3.3 Multi-objective Mode

In this Section, the results of the MO problem are presented. This approach constitutes a deep analysis on the influence of the different parameters of the optimization problem, and thus the potential of the proposed emplacement for the MHPP can be evaluated. The results of the MO approach consist in the Pareto front [60], which is formed by a set of non-dominated solutions, that correspond to optimal combinations of the objective values. In this case, the economic objective of minimizing the cost of the MHPP and the maximization of the power generated are combined. The Pareto front is represented in Fig. 11.

Given the difficulty of interpreting its morphology, due to its three-dimensional nature, an interpolation surface is proposed in Fig. 12, which is developed by means of a tessellation for a better understanding.

In sight of the morphology of the Pareto front shown in Fig. 12, some comments can be made. For this, a simple scheme of the Pareto front is drawn in Fig. 13. Using this scheme, the result of fixing each of the three objectives (cost of the penstock, cost of the distribution line, and power) can be evaluated.

First, fixing the cost of the distribution line, C_e, (see Fig. 14 top) implies a maximum reachable power, which can be increased if the cost of the distribution line is also increased. This is reasonable, as fixing the cost of the distribution grid translates into fixing the location of the dam. The higher this cost is, the further the dam is placed from the community, and the larger range of domain is thus available.

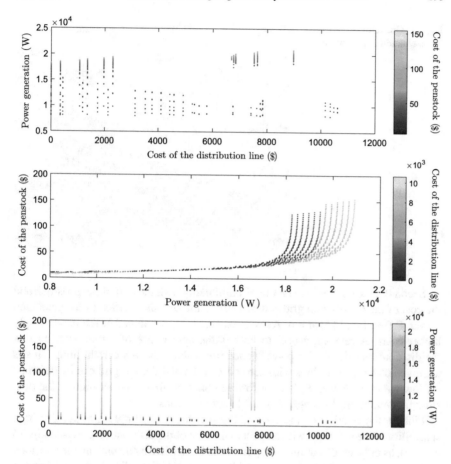

Fig. 11 Representation of the Pareto Front

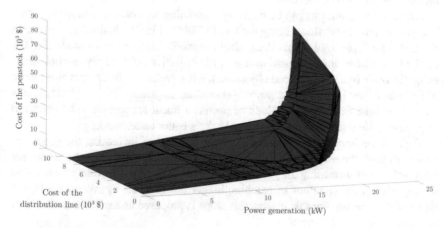

Fig. 12 Interpolation of the Pareto Front

Fig. 13 Qualitative
representation of the Pareto
front

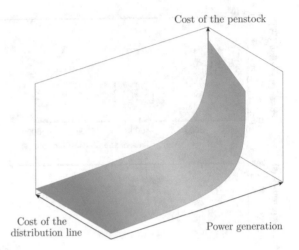

Secondly, the competitive nature of minimizing both the cost of the penstock and the cost of the distribution grid can be verified through the fixation of the generated power P (see Fig. 14 middle). As was expected, higher values of the power imply higher costs. A relevant aspect to note is the appearance of a minimum cost of the distribution grid, C_e for power generation values above a certain limit. This is understandable, as solutions with zero cost of the distribution grid can be obtained for low values of power P, but not for high ones, as fixing the location of the dam strongly conditions the range of possible combinations.

Finally, the fixation of the cost of the penstock, C_p, imply a extremely high sensibility of the solutions with respect to the cost of the distribution grid (see Fig. 14 bottom), as only small variations in power can be reached by displacing the penstock along the terrain, with the consequent high variances of the distribution grid due to moving the location of the dam.

An interesting analysis can be made by combining the costs related to the distribution grid and the penstock, being the Pareto front in Fig. 15 obtained as a result. In sight of the morphology of this Pareto front, some comments can be made. First, it can be seen that an improvement in one of the objectives necessarily implies worsening the other (counterbalanced objectives), what evidences the competitive nature of the cost of the plant and the power generation. Regarding the morphology, it is relevant to note that, for low values of power, a linear tendency can be observed (represented by a red dashed line in Fig. 15 for a better understanding).

This can be interpreted as a constant marginal cost of increasing the generated power. Indeed, the slope of this tendency, $r = \$656/kW$, represents the increment marginal cost of increasing the generated power in 1 kW. It can be easily understood the capacity of the solutions of reaching higher power levels by covering a higher range of the domain. For this reason, this slope is proposed as a reliable indicator of potential of the emplacement.

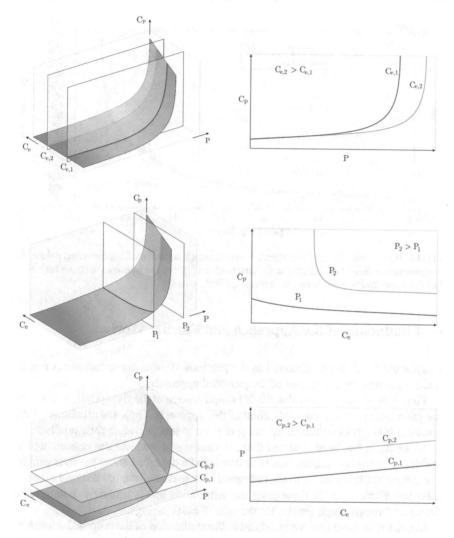

Fig. 14 Qualitative analysis of the morphology of the Pareto front

In addition, the abrupt change of this linear tendency which is observed for high values of power can also be understood by considering that the solutions can saturate the domain. For these solutions, the penstock covers the entire domain, and thus only small improvements in power can be reached by either modifying the nodes distribution and increasing the pipe diameter, which strongly affects the cost.

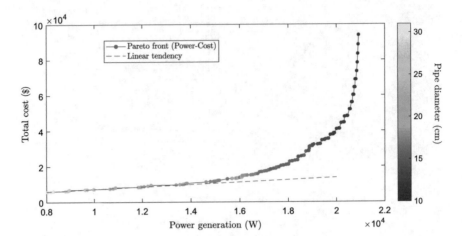

Fig. 15 Pareto front Power-Cost, electrical and penstock costs are added (coloured points and continuous blue line). The diameter of the pipe has been represented by color (see color bar). Note that the linear tendency has been also represented (red dashed)

6 Limitations of the Approach and Further Work

Despite the good results obtained in the previous section, some comments can be made regarding the limitations of the proposed approach.

First, it is relevant to note that the 2D simplification of the river profile constitutes the main limitation for the application of this approach. Thus, the reliability of the results is strongly conditioned by its application to low curvature river profiles.

Secondly, it has been assumed that the village and the river are near enough to consider that the distribution line is desirable to be deployed along the river profile. For this reason, this approach is not adequate if there is a substantial distance between these two elements, as in these cases the distribution line is generally desired to be deployed through rough terrain, for the sake of costs savings.

Considering these two issues related to the application of the proposed approach, the extension of this study to a 3D domain is proposed as a further work, with the aim of improving the modeling of the distribution line and providing a reliable model of the deployment of the penstock for rivers with non-negligible curvature.

7 Conclusions

The main conclusions of the work presented in this chapter can be summarized as follows:

- A model of a MHPP has been developed, allowing the study of the influence of the different design parameters on the performance and the costs.

- A GA has been developed to address the optimization problem of finding the most suitable layout of the MHPP.
- The proposed approach has been successfully applied to a real case in a remote village in Honduras, where a set of topographic data has been used as input.
- A single-objective mode (total cost minimization) has been first used to address the optimization problem. An optimal solution with cost \$9942.5 has been obtained, demonstrating the capability of the algorithm to address complex non-linear problem formulations with good convergence rates.
- The configuration $p_{cx} = 0.5$, $p_{mut} = 0.5$ has demonstrated the best performance, providing the lowest mean fitness (\$10572) and standard deviation (\$1665.4) in the final population.
- A multi-objective mode (power maximization, penstock cost minimization and distribution line minimization) has additionally been applied to the optimization problem, being the Pareto front determined. This has provided a deep study of the potential of the emplacement. The marginal cost of increasing the generated power of the MHPP has been determined in \$656 per additional kWatt installed.

Acknowledgements This research has been partially funded by the University of Seville under the contract "Contratos de acceso al Sistema Español de Ciencia, Tecnología e Innovación para el desarrollo del programa propio de I+D+i de la Universidad de Sevilla" of D. G. Reina.

References

1. Nejat, Payam, Fatemeh Jomehzadeh, Mohammad Mahdi Taheri, Mohammad Gohari, and Muhd Zaimi Abd Majid. 2015. A global review of energy consumption, co2 emissions and policy in the residential sector (with an overview of the top ten co2 emitting countries). *Renewable and Sustainable Energy Reviews* 43: 843–862.
2. International Energy Agency. 2017. WEO-2017 Special Report: Energy Access Outlook.
3. Pereira, Marcio Giannini, José Antonio Sena, Marcos Aurélio Vasconcelos Freitas, and Neilton Fidelis Da Silva. 2011. Evaluation of the impact of access to electricity: A comparative analysis of South Africa, China, India and Brazil. *Renewable and Sustainable Energy Reviews* 15 (3): 1427–1441.
4. OECD. 2016. *Linking renewable energy to rural development, OECD green growth studies.* OECD Publishing.
5. Krewitt, Wolfram, Sonja Simon, Wina Graus, Sven Teskec, Arthouros Zervos, and Oliver Schafer. 2008. The 2 degrees c scenario-a sustainable world energy perspective (vol 35, pg 4969, 2007). *Energy Policy* 36 (1): 494–494.
6. Luderer, Gunnar, Volker Krey, Katherine Calvin, James Merrick, Silvana Mima, Robert Pietzcker, Jasper Van Vliet, and Kenichi Wada. 2014. The role of renewable energy in climate stabilization: Results from the EMF27 scenarios. *Climatic Change* 123 (3–4): 427–441.
7. United Nations Environment Programme. 2009. Renewables global status report: 2009 update. *Internet.* http://www.unep.fr/shared/docs/publications/RE_GSR_2009_Update.pdf.
8. Kanase-Patil, A.B., R.P. Saini, and M.P. Sharma. 2010. Integrated renewable energy systems for off grid rural electrification of remote area. *Renewable Energy* 35 (6): 1342–1349.
9. Bugaje, I.M. 2006. Renewable energy for sustainable development in Africa: A review. *Renewable and Sustainable Energy Reviews* 10 (6): 603–612.
10. Sahoo, Sarat Kumar. 2016. Renewable and sustainable energy reviews solar photovoltaic energy progress in India: A review. *Renewable and Sustainable Energy Reviews* 59: 927–939.

11. Saheb-Koussa, D., Mourad Haddadi, and Maiouf Belhamel. 2009. Economic and technical study of a hybrid system (wind–photovoltaic–diesel) for rural electrification in Algeria. *Applied Energy* 86 (7–8): 1024–1030.
12. Mohammed, Y.S., A.S. Mokhtar, Naomi Bashir, and R. Saidur. 2013. An overview of agricultural biomass for decentralized rural energy in Ghana. *Renewable and Sustainable Energy Reviews* 20:15–25.
13. Sachdev, Hira Singh, Ashok Kumar Akella, and Niranjan Kumar. 2015. Analysis and evaluation of small hydropower plants: A bibliographical survey. *Renewable and Sustainable Energy Reviews* 51: 1013–1022.
14. Kaldellis, J., and K. Kavadias. 2000. *Laboratory applications of renewable energy sources*. Athens: Stamoulis.
15. ESMAP. 2017. Technical and economic assessment of off-grid. Technical report, 12.
16. Muhammad Indra, al Irsyad, Anthony Halog, and Rabindra Nepal. 2019. Renewable energy projections for climate change mitigation: An analysis of uncertainty and errors. *Renewable Energy* 130: 536–546.
17. Jawahar, C.P., and Prawin Angel Michael. 2017. A review on turbines for micro hydro power plant. *Renewable and Sustainable Energy Reviews* 72: 882–887.
18. Khurana, S., and Anoop Kumar. 2011. Small hydro power, a review. *International Journal of Thermal Technologies* 1(1): 107–110.
19. Mandelli, Stefano, Jacopo Barbieri, Riccardo Mereu, and Emanuela Colombo. 2016. Off-grid systems for rural electrification in developing countries: Definitions, classification and a comprehensive literature review. *Renewable and Sustainable Energy Reviews* 58: 1621–1646.
20. Carravetta A., Derakhshan Houreh S., and Ramos H.M. 2018. *Pumps as turbines. Springer tracts in mechanical engineering*. Springer.
21. Cobb, Bryan R., and Kendra V. Sharp. 2013. Impulse (turgo and pelton) turbine performance characteristics and their impact on pico-hydro installations. *Renewable Energy* 50: 959–964.
22. Sangal, Saurabh, Arpit Garg, and Dinesh Kumar. 2013. Review of optimal selection of turbines for hydroelectric projects. *International Journal of Emerging Technology and Advance Engineering* 3: 424–430.
23. Herschy, Reginald W. 2014. *Streamflow Measurement*, vol. 3.
24. Thake, J. 2000. *Micro-hydro Pelton Turbine Manual: Design, manufacture and installation for small-scale hydropower*.
25. Mishra, Sachin, S.K. Singal, and D.K. Khatod. 2011. Optimal installation of small hydropower plant. A review. *Renewable and Sustainable Energy Reviews* 15 (8): 3862–3869.
26. Elbatran, A.H., O.B. Yaakob, Yasser M. Ahmed, and H.M. Shabara. 2015. Operation, performance and economic analysis of low head micro-hydropower turbines for rural and remote areas: A review. *Renewable and Sustainable Energy Reviews* 43: 40–50.
27. Haddad, Omid Bozorg, Mahdi Moradi-Jalal, and Miguel A. Marino. 2011. Design–operation optimisation of run-of-river power plants. In *Proceedings of the institution of civil engineers-water management*, vol. 164, 463–475. Thomas Telford Ltd.
28. Anagnostopoulos, John S., and Dimitris E. Papantonis. 2007. Optimal sizing of a run-of-river small hydropower plant. *Energy Conversion and Management* 48 (10): 2663–2670.
29. Alexander, K.V., and E.P. Giddens. 2008. Optimum penstocks for low head microhydro schemes. *Renewable Energy* 33 (3): 507–519.
30. Tapia, A., P. Millán, and F. Gómez-Estern. 2018. Integer programming to optimize micro-hydro power plants for generic river profiles. *Renewable Energy* 126: 905–914.
31. Gingold, P.R. 1981. The optimum size of small run-of-river plants. *International Water Power and Dam Construction* 33 (11).
32. Marliansyah, Romy, Dwini Normayulisa Putri, Andy Khootama, and Heri Hermansyah. 2018. Optimization potential analysis of micro-hydro power plant (MHPP) from river with low head. *Energy Procedia* 153: 74–79.
33. Iqbal, M., M. Azam, M. Naeem, A.S. Khwaja, and A. Anpalagan. 2014. Optimization classification, algorithms and tools for renewable energy: A review. *Renewable and Sustainable Energy Reviews* 39: 640–654.

34. Banos, Raul, Francisco Manzano-Agugliaro, F.G. Montoya, Consolacion Gil, Alfredo Alcayde, and Julio Gómez. 2011. Optimization methods applied to renewable and sustainable energy: A review. *Renewable and Sustainable Energy Reviews* 15 (4): 1753–1766.
35. Singal, S.K., R.P. Saini, and C.S. Raghuvanshi. 2010. Analysis for cost estimation of low head run-of-river small hydropower schemes. *Energy for Sustainable Development* 14 (2): 117–126.
36. Mohamad, Hasmaini, Hazlie Mokhlis, Hew Wooi Ping, et al. 2011. A review on islanding operation and control for distribution network connected with small hydro power plant. *Renewable and Sustainable Energy Reviews* 15 (8): 3952–3962.
37. Nand Kishor, R.P. Saini, and S.P. Singh. 2007. A review on hydropower plant models and control. *Renewable and Sustainable Energy Reviews* 11 (5): 776–796.
38. Leon, Arturo S., and Ling Zhu. 2014. A dimensional analysis for determining optimal discharge and penstock diameter in impulse and reaction water turbines. *Renewable Energy* 71: 609–615.
39. Basso, S., and G. Botter. 2012. Streamflow variability and optimal capacity of run-of-river hydropower plants. *Water Resources Research* 48 (10).
40. Yildiz, Veysel, and Jasper A. Vrugt. 2019. A toolbox for the optimal design of run-of-river hydropower plants. *Environmental Modelling & Software* 111: 134–152.
41. Yoo, Ju-Hwan. 2009. Maximization of hydropower generation through the application of a linear programming model. *Journal of Hydrology* 376 (1–2): 182–187.
42. Das, Himansu, Ajay Kumar Jena, Janmenjoy Nayak, Bighnaraj Naik, and H.S. Behera. 2015. A novel PSO based back propagation learning-MLP (PSO-BP-MLP) for classification. In *Computational intelligence in data mining*, vol. 2, 461–471. Springer.
43. Nayak, Janmenjoy, Bighnaraj Naik, A.K. Jena, Rabindra K. Barik, and Himansu Das. 2018. Nature inspired optimizations in cloud computing: Applications and challenges. In *Cloud computing for optimization: Foundations, applications, and challenges*, 1–26. Springer.
44. Alvarado-Barrios, Lázaro, A. Rodríguez del Nozal, A. Tapia, José Luis Martínez-Ramos, and D.G. Reina. 2019. An evolutionary computational approach for the problem of unit commitment and economic dispatch in microgrids under several operation modes. *Energies* 12 (11): 2143.
45. Anagnostopoulos, John S., and Dimitrios E. Papantonis. 2007. Flow modeling and runner design optimization in turgo water turbines. *International Journal of Mechanical, Aerospace, Industrial and Mechatronics Engineering*, 1 (4): 204–209.
46. Ehteram, Mohammad, Hojat Karami, Sayed-Farhad Mousavi, Saeed Farzin, and Ozgur Kisi. 2018. Evaluation of contemporary evolutionary algorithms for optimization in reservoir operation and water supply. *Journal of Water Supply: Research and Technology-Aqua* 67 (1): 54–67.
47. Tapia, A., D.G. Reina, and P. Millán. 2019. An evolutionary computational approach for designing micro hydro power plants. *Energies* 12 (5): 878.
48. Holland, John H. 1984. *Genetic algorithms and adaptation*. USA: Springer.
49. Itō, H., and K. Imai. 1973. Energy losses at 90 pipe junctions. *Journal of the Hydraulics Division* 99 (9): 1353–1368.
50. Dasgupta, Dipankar, and Zbigniew Michalewicz. 2013. *Evolutionary algorithms in engineering applications*. Springer Science & Business Media.
51. Reina, D.G., T. Camp, A. Munjal, and S.L. Toral. 2018. Evolutionary deployment and local search-based movements of 0th responders in disaster scenarios. *Future Generation Computer Systems* 88: 61–78.
52. Arzamendia, Mario, Derlis Gregor, Daniel Gutierrez Reina, and Sergio Luis Toral. 2017. An evolutionary approach to constrained path planning of an autonomous surface vehicle for maximizing the covered area of Ypacarai lake. *Soft Computing* 1–12.
53. Gutiérrez-Reina, Daniel, Vishal Sharma, Ilsun You, and Sergio Toral. 2018. Dissimilarity metric based on local neighboring information and genetic programming for data dissemination in vehicular ad hoc networks (vanets). *Sensors* 18 (7): 2320.
54. Deb, Kalyanmoy, Amrit Pratap, Sameer Agarwal, and TAMT Meyarivan. 2002. A fast and elitist multiobjective genetic algorithm: Nsga-ii. *IEEE transactions on evolutionary computation* 6 (2): 182–197.
55. Ter-Sarkisov, Aram, and Stephen Marsland. 2011. Convergence properties of two $(\mu + \lambda)$ evolutionary algorithms on onemax and royal roads test functions. arXiv:1108.4080.

56. Luke, Sean. 2009. *Essentials of metaheuristics*, vol. 113. Lulu Raleigh.
57. Reina, D.G., J.M. León-Coca, S.L. Toral, E. Asimakopoulou, F. Barrero, P. Norrington, and N. Bessis. 2014. Multi-objective performance optimization of a probabilistic similarity/dissimilarity-based broadcasting scheme for mobile ad hoc networks in disaster response scenarios. *Soft Computing* 18 (9): 1745–1756.
58. Gutiérrez, D. 2018. Evolutionary MHPP. https://github.com/Dany503/Evolutionary-MHPP.
59. Fortin, Félix-Antoine, François-Michel De Rainville, Marc-André Gardner, Marc Parizeau, and Christian Gagné. 2012. Deap: Evolutionary algorithms made easy. *Journal of Machine Learning Research* 13 (Jul): 2171–2175.
60. Reina, D.G., R.I. Ciobanu, S.L. Toral, and C. Dobre. 2016. A multi-objective optimization of data dissemination in delay tolerant networks. *Expert Systems with Applications* 57 (C): 178–191.

Performance Evaluation of Different Machine Learning Methods and Deep-Learning Based Convolutional Neural Network for Health Decision Making

Abhaya Kumar Sahoo, Chittaranjan Pradhan and Himansu Das

Abstract Now-a-days modern technology is used for health management and diagnostic strategy in the health sector. Machine learning usually helps in decision making for health issues using different models. Classification and prediction of disease are easily known with the help of machine learning techniques. The machine learning technique can be applied in various applications such as image segmentation, fraud detection, pattern recognition and disease prediction, etc. In the today's world, maximum people are suffering from diabetes. The glucose factor in the blood is the main component of diabetes. Fluctuation of blood glucose level leads to diabetes. To predict the diabetes disease, machine learning and deep learning play major role which uses probability, statistics and neural network concepts, etc. Deep learning is the part of machine learning which uses different layers of neural network that decide classification and prediction of disease. In this chapter, we study and compare among different machine learning algorithms and deep neural networks for diabetes disease prediction, by measuring performance. The experiment results prove that convolution neural network based deep learning method provides the highest accuracy than other machine learning algorithms.

Keywords Convolutional neural network · Deep learning · Diabetes disease prediction · Machine learning · Performance evaluation

1 Introduction

Now-a-days, maximum people are suffering from diabetes. It is very important to identify this disease among patients. Classification is a technique which helps to classify data into different classes based on classifier. So Diabetes disease prediction requires classification method to predict disease by using different models. Diabetes disease increases glucose level in the blood which affects in production of insulin hormone. This disease causes malfunction of pancreatic beta cell and also leads to

A. K. Sahoo (✉) · C. Pradhan · H. Das
School of Computer Engineering, KIIT Deemed to be University, Bhubaneswar, Odisha, India
e-mail: abhayakumarsahoo2012@gmail.com

© Springer Nature Switzerland AG 2020
M. Rout et al. (eds.), *Nature Inspired Computing for Data Science*,
Studies in Computational Intelligence 871,
https://doi.org/10.1007/978-3-030-33820-6_8

impact on different parts of the body like heart, kidney, nerve, foot, eye etc. [1]. For classification of disease, we use machine learning and deep learning method.

Machine learning is an emerging field which is applied to large data sets to extract hidden concepts and relationship among attributes. It is in the form of algorithm, i.e. used by model to predict the outcome. It is very difficult to handle and process the large volume of data by human. So machine learning plays a main role to predict healthcare outcomes with cost minimization and high quality [2]. The machine learning algorithms are mainly based on probability-based, tree based and rule-based etc. Huge amount of data collected from different sources, are used in data-preprocessing phase. In this phase, dimension of data is reduced by removing irrelevant data. As the volume of data is increased, the system is unable to make a decision. So different algorithms should be developed in such a way that useful patterns or hidden knowledge can be extracted from historical data. After developing the model using machine learning algorithms, the model is tested using a test dataset to find the accuracy of the model. The algorithm used in the model can be again optimized by considering some parameters or rules. Basically, machine learning is used in the field of data classification, pattern recognition and prediction. The different applications such as face detection, disease prediction, fraud detection, email filtering and traffic management, etc. use the machine learning concepts [3]. Deep learning is the part of machine learning, which uses supervised and unsupervised techniques for feature classification. The different components of deep learning are used in the area of image segmentation, disease prediction and recommender system such as convolution neural networks, auto-encoders and restricted-Boltzmann machines etc.

The main contribution of this chapter is to compare different machine learning and deep learning techniques for diabetes prediction and measure the performance of all methods in the term of accuracy. In this context, Convolution neural network (CNN) gives better performance as per compared to all methods. The rest of the chapter is divided into following sections. Section 2 presents an explanation of machine learning, deep learning and diabetes disease. Section 3 shows various machine learning approaches. Section 4 explains proposed CNN method in disease prediction. Section 5 shows the experimental result. Finally, Sect. 6 concludes the chapter along with future work.

2   Literature Survey

In this paper, we give more focus on various approaches of machine learning and deep learning. Different approaches of machine learning and deep learning are explained below.

A.   *Machine Learning*

Machine learning [4–10] is a technique by which we build a model and that model decides and predicts the outcomes. Before developing model, first we load the dataset, and then data pre processing is done on the input data. Different classifiers are used to train the input data. After training the data, the model is developed, which tests the

Fig. 1 Steps involved in
machine learning process

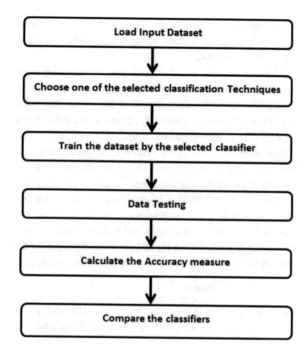

test dataset. At last we calculate the accuracy of model. In this way we can compare
the accuracy among all classifiers. Figure 1 shows the steps involved in machine
learning technique, which predicts the possible outcomes [11–13].

Machine learning is based on three methods such as supervised, unsupervised and
reinforcement. In Supervised learning method, class label is defined. The model is
trained to test the data and predict the outcomes based on a predefined class label.
In unsupervised learning method, class label is not defined. After developing the
model, model is used to find patterns and relationships in the dataset. In reinforcement
learning, it gives the ability of an agent to interact with the environment and find the
best outcome using hit and trial method. Different machine learning algorithms such
as KNN, logistic regression, decision tree, random forest, support vector machine
and multi layer perceptron etc. are used in diabetes disease prediction [2, 14–16].

B. *Deep Learning*

Deep learning is a subset of machine learning methods based on learning data
representations. Neural network is the building block in deep learning, which is
inspired by information processing and distributed communication nodes in biological systems. Deep learning has different components such as Auto-encoder
(AE), Restricted Boltzmann Machine (RBM), Convolution Neural Network (CNN),
Multilayer Perceptron (MLP) and Recurrent Neural Network (RNN) etc. [17–20].

Multilayer perceptron with auto encoder is a feed forward neural network that
contains many hidden layers. Each layer contains perceptrons which use arbitrary
activation function. Auto-encoder is an unsupervised learning model that tries to

regenerate its input data in the output. This neural network uses back propagation technique which calculates the gradient of the error function with respect to the neural network's weights.

Auto-encoders are the main part of learning representation which encode the input and converts the large vectors into small vectors, which is depicted in Fig. 2. The small vectors record the most significant features of vectors which help in data compression, dimensionality reduction and data reconstruction.

CNN is a feed-forward neural network, which consists of convolution layers. These convolution layers capture the local and global feature that leads to the enhancement of accuracy. The RBM consists of two layer neural network such as visible layer and a hidden layer. There is no intra layer communication among the visible layer and hidden layer. This RBM uses a gradient descent approximation method that helps in extracting features and attributes. Figure 3 shows the RBM network consisting of visible and hidden units VI and Hg, respectively and strength of connections between these layers are denoted by a matrix of weights W, which is shown in Fig. 3. The performance of the model depends on no. of hidden layers.

C. *Diabetes Disease*

Diabetes mellitus is a disease in which maltose level does not maintain in the blood. Due to high maltose level in blood, cells inhibit the production of insulin. High

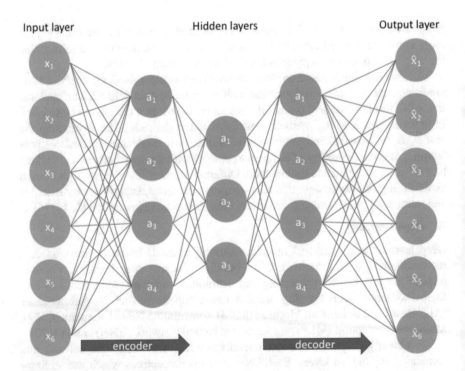

Fig. 2 Auto-encoder neural network

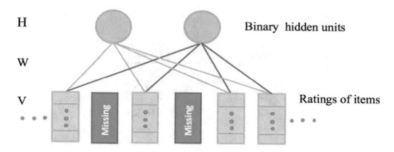

Fig. 3 RBM neural network

maltose level in blood leads to tiredness, headache and itching skin, etc. [21–23]. Type-2 diabetes is a type of diabetes mellitus in which insulin is not properly produced in the body, whereas Type-1 diabetes is a disease in which pancreas is damaged that does not produce insulin [24–26].

3 Different Machine Learning Approaches

Different machine learning classifiers are used to build the model which predicts the diabetes disease. These classifiers are explained as follows.

A. *K-Nearest Neighbor (KNN) Method*

K-Nearest Neighbors is a classification algorithm which is based on supervised learning method. In this algorithm, the whole dataset is used in training phase, i.e. required to build the model. KNN searches the entire training dataset and find k-most similar instances. Most similar instance data provide the prediction result for unseen data instance [11]. This classifier is shown in Fig. 4.

KNN algorithm is described as follows:

Algorithm: KNN Algorithm
Input: Data set
Output: Class label of test data
1: Import the dataset
2: Define the value of k
3: For finding the predicted class, start iteration from 1 to the total number of training data points
Do
    i. Find the distance between test data and training data points of each row.
     Euclidean distance is used as a distance metric to find the distance among test data point and training data point.
    ii. Arrange the calculated distances in ascending order based on distance values.
    iii. Find top k rows from the sorted array.
    iv. Find the most frequent class of these rows.
    v. Return the predicted class.
4: End for

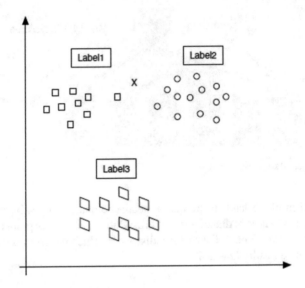

Fig. 4 KNN classifier

B. *Logistic Regression*

Logistic regression is a supervised learning method which is used to predict outcomes. When a class label is binary categorical, this method is used. The binary classification problem can be easily solved by this method. According to logistic regression method, when X value is known, Y value will be predicted. Generally, this categorical variable Y contains two classes, which is shown in Fig. 5. If Y contains more than 2 classes, then it is called multiclass classification problem. The main advantage of logistic regression is to compute prediction probability score of an event [11, 27].

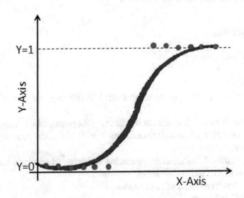

Fig. 5 Logistic regression

C. *Decision Tree Method*

The decision tree is a supervised learning based classification algorithm. This algorithm uses a treelike model for taking decision and predicting the outcomes. Decision tree starts with root named as splitting attribute. This root like single node branches into possible outcomes and each outcome leads into additional nodes that divide into other possibilities [27]. This method is represented in Fig. 6.

D. *Support Vector Machine Method*

Support vector machine is classifier which is based on a separating hyperplane. This method gives an optimal hyperplane by which training data are categorized. The hyperplane in two dimensional spaces is a line which divides the plane into two parts. The two parts define two separate classes, which is shown in Fig. 7. The main idea behind SVM is to draw a decision boundary which separates the dataset into two distinct classes [14].

E. *Random Forest Method*

Random forest method is a supervised learning method used to solve both classification and regression problem. This method creates a forest with several trees. This method is represented in Fig. 8. The random forest classifier tells that if the number of trees in forest is higher, then the accuracy of the model will be higher [11].

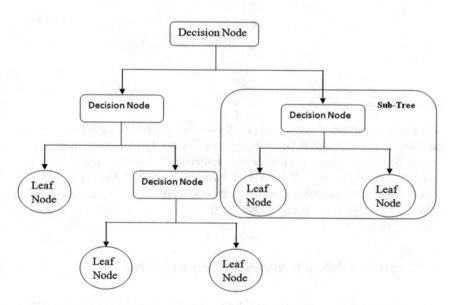

Fig. 6 Decision tree classifier

Fig. 7 Support vector
machine

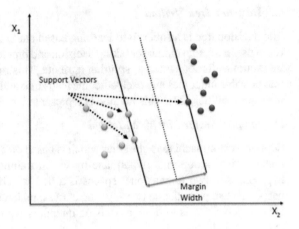

Fig. 8 Random forest
classifier

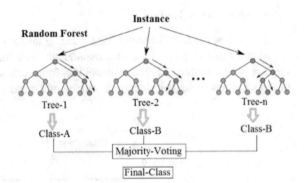

F. *Multi Layer Perceptron*

Multilayer perceptron is an artificial neural network, which consists of more than one perceptron. The perceptrons in input layer receive the signal and output layer tales the responsibility to make the decision or prediction. The hidden layers behave as a computational engine in MLP. The MPL is based on supervised learning method used to solve classification problem, which is shown in Fig. 9. The model uses training parameters, weights and biases minimize the error. Back propagation is used to minimize the error [11].

4 Proposed CNN in Diabetes Disease Prediction

Convolution Neural Network is a feed forward neural network consisting of convolution layers which captures the local and global features of the input. CNN helps in modeling of sequential data. The basic units of CNN are conventional layer, activation function, pooling layer and fully connected layer. The convolution layer is used as feature extractor i.e. described in Eq. 1.

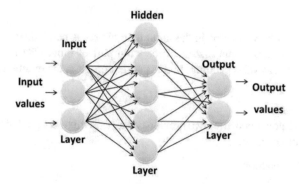

Fig. 9 Multilayer perceptron

$$f_c(x) = act\left(\sum_{i,j}^{n} W_{c(n-i)(n-j)}x_{ij} + b_c\right) \tag{1}$$

where W_c and n represent the weight matrix and size of perception field. b_c denotes the bias vector and act(x) is the activation function which makes the neural network able to approximate any nonlinear function. Figure 10 depicts the different phases of CNN layer.

Algorithm

Step1: Convolution Operation
Convolution layer is used to extract features from input data. The input data is converted into a number of small squares of data. This convolution layer uses convolution operation which keeps the relationship among data values by learning features. Convolution operation is a mathematical expression that uses two inputs as data matrix and 3 × 3 filters.

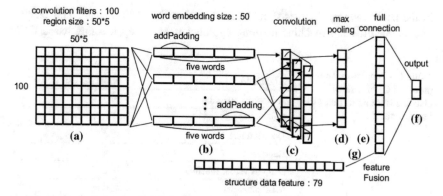

Fig. 10 Convolution neural network in diabetes disease prediction

Step2: Pooling
Padding the inputs with zeros in the input data matrix is performed when the filter is not fitted with input data matrix. If the valid part of input data is not lost, then padding is called as valid padding. In this pooling process, the rectified linear unit is used.

Step3: Flattening
Feature map is generated after two previous phases. In this step, pooled featured map is flattened into a column.

Step4: Full Connection
Full connected layer is created.

5   Experimental Result Analysis

Data Set
Pima Indians Diabetes Data Set describes the medical records for Pima Indians. This dataset contains 2000 rows and 9 columns. In this data set, all patients are female having at least 21 years old. The dataset is used to find whether the patient is diabetic patient or not.

Here we compare all machine learning methods with proposed CNN method based on diabetes dataset. We get more accuracy in proposed CNN than random forest method which is shown in Table 1.

The Fig. 11 shows the comparison graph among proposed CNN and all machine learning methods. This graph signifies that CNN based deep learning method gives better result in term of accuracy than ML methods.

6   Conclusion

In the health sector, diabetes should be detected at the earliest stage. This chapter tells about different machine learning approaches and deep learning based CNN

Table 1 Comparison among proposed CNN with all machine learning methods to find accuracy

Algorithm	Accuracy (%)
K-Nearest Neighbors	79
Logistic regression	77.8
Decision tree	82.6
Random forest	92.4
Support vector machine	78.2
MLP classifier	82.4
Proposed CNN	93.2

Fig. 11 Comparison graph between proposed CNN and all ML methods

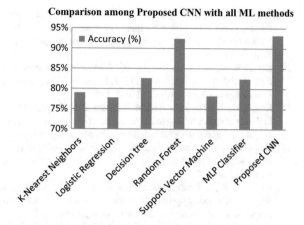

method for diabetes disease prediction. In the experiment results, we conclude that CNN based deep learning technique provides accuracy of 93.2%, which is better than a random forest classification algorithm. In the future, we will try to use a hybrid approach to get better accuracy than existing methods.

References

1. Sneha, N., and T. Gangil. 2019. Analysis of diabetes mellitus for early prediction using optimal features selection. *Journal of Big Data* 6 (1): 13.
2. Doupe, P., J. Faghmous, and S. Basu. Machine learning for health services researchers. *Value in Health*.
3. Kaur, H., and V. Kumari. 2018. Predictive modelling and analytics for diabetes using a machine learning approach. *Applied Computing and Informatics*.
4. Das, H., B. Naik, and H.S. Behera. 2018. Classification of diabetes mellitus disease (DMD): a data mining (DM) approach. In *Progress in computing, analytics and networking*, 539–549. Singapore: Springer.
5. Sahani, R., C. Rout, J.C. Badajena, A.K. Jena, and H. Das. 2018. Classification of intrusion detection using data mining techniques. In *Progress in computing, analytics and networking*, 753–764. Singapore: Springer.
6. Das, H., A.K. Jena, J. Nayak, B. Naik, and H.S. Behera. 2015. A novel PSO based back propagation learning-MLP (PSO-BP-MLP) for classification. In *Computational intelligence in data mining*, vol. 2, 461–471. New Delhi: Springer.
7. Pradhan, C., H. Das, B. Naik, and N. Dey. 2018. *Handbook of research on information security in biomedical signal processing*, 1–414. Hershey, PA: IGI Global. https://doi.org/10.4018/978-1-5225-5152-2.
8. Pattnaik, P.K., S.S. Rautaray, H. Das, and J. Nayak (eds.). 2018. Progress in computing, analytics and networking. In *Proceedings of ICCAN 2017*, vol. 710. Springer.
9. Nayak, J., B. Naik, A.K. Jena, R.K. Barik, and H. Das. 2018. Nature inspired optimizations in cloud computing: applications and challenges. In *Cloud computing for optimization: foundations, applications, and challenges*, 1–26. Cham: Springer.
10. Mishra, B.B., S. Dehuri, B.K. Panigrahi, A.K. Nayak, B.S.P. Mishra, and H. Das. 2018. *Computational intelligence in sensor networks*, vol. 776. Studies in Computational Intelligence. Springer.

11. Kanchan, B.D., and M.M. Kishor. 2016. Study of machine learning algorithms for special disease prediction using principal of component analysis. In *2016 international conference on global trends in signal processing, information computing and communication (ICGTSPICC)*, 5–10. IEEE.
12. Khalil, R.M., and A. Al-Jumaily. 2017. Machine learning based prediction of depression among type 2 diabetic patients. In *2017 12th international conference on intelligent systems and knowledge engineering (ISKE)*, 1–5. IEEE.
13. Dey, S.K., A. Hossain, and M.M. Rahman. 2018. Implementation of a web application to predict diabetes disease: an approach using machine learning algorithm. In *2018 21st international conference of computer and information technology (ICCIT)*. 1–5. IEEE.
14. Barakat, N., A.P. Bradley, and M.N.H. Barakat. 2010. Intelligible support vector machines for diagnosis of diabetes mellitus. *IEEE Transactions on Information Technology in Biomedicine* 14 (4): 1114–1120.
15. Sahoo, A.K., C. Pradhan, and B.S.P. Mishra. 2019. SVD based privacy preserving recommendation model using optimized hybrid item-based collaborative filtering. In *2019 international conference on communication and signal processing (ICCSP)*, 0294–0298. IEEE.
16. Sahoo, A.K., S. Mallik, C. Pradhan, B.S.P. Mishra, R.K. Barik, and H. Das. 2019. Intelligence-based health recommendation system using big data analytics. In *Big data analytics for intelligent healthcare management*, 227–246. Academic Press.
17. Jan, B., H. Farman, M. Khan, M. Imran, I.U. Islam, A. Ahmad, and G. Jeon. 2017. Deep learning in big data analytics: a comparative study. *Computers & Electrical Engineering*.
18. Zhao, R., R. Yan, Z. Chen, K. Mao, P. Wang, and R.X. Gao. 2019. Deep learning and its applications to machine health monitoring. *Mechanical Systems and Signal Processing* 115: 213–237.
19. Miotto, R., F. Wang, S. Wang, X. Jiang, and J.T. Dudley. 2017. Deep learning for healthcare: review, opportunities and challenges. *Briefings in Bioinformatics* 19 (6): 1236–1246.
20. Sahoo, A.K., C. Pradhan, R.K. Barik, and H. Dubey. 2019. DeepReco: deep learning based health recommender system using collaborative filtering. *Computation* 7 (2): 25.
21. Solanki, J.D., S.D. Basida, H.B. Mehta, S.J. Panjwani, and B.P. Gadhavi. 2017. Comparative study of cardiac autonomic status by heart rate variability between under-treatment normotensive and hypertensive known type 2 diabetics. *Indian Heart Journal* 69 (1): 52–56.
22. Baiju, B.V., and D.J. Aravindhar. 2019. Disease influence measure based diabetic prediction with medical data set using data mining. In *2019 1st international conference on innovations in information and communication technology (ICIICT)*, 1–6. IEEE.
23. Undre, P., H. Kaur, and P. Patil. 2015. Improvement in prediction rate and accuracy of diabetic diagnosis system using fuzzy logic hybrid combination. In *2015 international conference on pervasive computing (ICPC)*, 1–4. IEEE.
24. Prasad, S.T., S. Sangavi, A. Deepa, F. Sairabanu, and R. Ragasudha. 2017. Diabetic data analysis in big data with predictive method. In *2017 international conference on algorithms, methodology, models and applications in emerging technologies (ICAMMAET)*, 1–4. IEEE.
25. Hammoudeh, A., G. Al-Naymat, I. Ghannam, and N. Obied. 2018. Predicting hospital readmission among diabetics using deep learning. *Procedia Computer Science* 141: 484–489.
26. Aliberti, A., I. Pupillo, S. Terna, E. Macii, S. Di Cataldo, E. Patti, and A. Acquaviva. 2019. *A multi-patient data driven approach to blood glucose prediction. IEEE Access.*
27. Sisodia, D., and D.S. Sisodia. 2018. Prediction of diabetes using classification algorithms. *Procedia Computer Science* 132: 1578–1585.

Clustering Bank Customer Complaints on Social Media for Analytical CRM via Multi-objective Particle Swarm Optimization

Rohit Gavval and Vadlamani Ravi

Abstract The ease of access to social media and its wide reach has made it the preferred platform for consumers to express their opinions and grievances regarding all types of products and services. As such, it serves as a fertile ground for organizations to deploy analytical customer relationship management (ACRM) and generate business insights of tremendous value. Of particular importance is identifying and quickly resolving customer complaints. If not redressed, these complaints could lead to customer churn and when addressed quickly and efficiently, they can double the profits. This automatic product-wise clustering of complaints helps in better sentiment analysis on products and services. In this paper, two variants of a novel multi-objective clustering algorithm are proposed with applications to sentiment analysis, an important area of analytical CRM in banking industry. The first variant, MOPSO-CD-Kmeans, employs Multi-objective Particle Swarm Optimization along with heuristics of K-means and the second variant, MOPSO-CD-SKmeans employs the same multi-objective particle swarm optimization along with the heuristics of Spherical K-means to find an optimal partitioning of the data. Two clustering criteria were considered as objective functions to be optimized to find a set of Pareto optimal solutions, and then the Silhouette Index was employed to determine the optimal number of clusters. The algorithm is then tested on bank based complaint datasets related to four Indian banks. Experiments indicate that MOPSO-CD-SKmeans is able to achieve promising results in terms of product-wise clustering of complaints and could outperform the first variant.

Keywords Clustering · Customer complaints · Social media · Analytical CRM · Multi-objective optimization · Particle swarm optimization

R. Gavval · V. Ravi (✉)
Center of Excellence in Analytics, Institute for Development and Research in Banking Technology (IDRBT), Castle Hills Road #1, Masab Tank, Hyderabad 500057, India
e-mail: vravi@idrbt.ac.in

R. Gavval
School of Computer and Information Sciences, University of Hyderabad, Hyderabad 500046, India

© Springer Nature Switzerland AG 2020
M. Rout et al. (eds.), *Nature Inspired Computing for Data Science*,
Studies in Computational Intelligence 871,
https://doi.org/10.1007/978-3-030-33820-6_9

1   Introduction

Customer relationship management (CRM) has proved to be highly impactful in changing the way businesses function. Several service industries like Retail, Healthcare, Travel, Hospitality, Telecom, Banking, Financial services and Insurance (BFSI) industry have embarked on ACRM journey to improve their growth. The BFSI industry is one of the largest users of ACRM owing to a variety of applications like credit scoring, market basket analysis, churn prediction, fraud detection, risk analytics, etc.

The CRM philosophy promotes a customer-centric approach for the functioning of enterprises and customer data is the fuel which drives these systems. With the ever-growing adoption and usage of social media by consumers, obtaining customer opinion data has never been easier. Organizations can analyze this 'gold mine' of data to learn a wide spectrum of information about their customers and take effective data-driven decisions. Particularly, a large number of consumers are increasingly turning to social media to express their concerns regarding various products and services. Organizations can leverage this data to expedite grievance redressal which can strengthen relationship with customers.

Customer complaints are available in the form of unstructured text data which cannot be processed using traditional analytical techniques. *Natural language processing* (NLP) along with *text mining* techniques contain useful tools to process and derive insights from text data. *Sentiment analysis* is the process which makes use of NLP and text mining techniques to understand the opinion in a given document available in written or spoken language. This line of study has been employed by several researchers to analyze customer feedback data. One popular task in sentiment analysis is *sentiment classification* in which customer feedback data is analyzed by automatically classifying them into negative or positive classes using supervised learning methods. Another popular task of sentiment analysis, called *opinion summarization* involves automatically generating a summary of the several documents related to a topic [1]. Before performing a full-fledged sentiment analysis of the complaints, automatically categorizing the complaints by product can enable the bank executives to greatly expedite the grievance redressal and improve customer satisfaction.

To meet this requirement, we propose a multi-objective clustering framework which can automatically segment a corpus of customer complaints into product-wise clusters. Clustering is a fundamental approach for data mining with a wide array of applications like identification of co-expressed genes in bioinformatics, image segmentation, recommendation systems, target marketing, spatial data analysis, information retrieval etc. In this study we propose a novel algorithm for clustering customer complaints. Clustering is an unsupervised learning technique and similar to other machine learning techniques, the performance of a clustering algorithm depends heavily on the quality and distribution of data. Traditional clustering algorithms like K-means [2] essentially optimize a single objective. These algorithms are susceptible to falling in local minima as they perform a greedy search over the error surface. Evolutionary and swarm based metaheuristics are population based optimization techniques which are adept at avoiding local minima and increase the chances of finding the global optima. Hence, researchers extensively employed evolutionary and swarm based optimization algorithms for the clustering task. As different datasets can have different shapes, optimizing one objective may not guarantee the best partitioning of a dataset. Like other real-world problems, good clustering relies on simultaneous optimization of multiple objectives. A large repertoire of cluster validity indices have been proposed in the literature which define the quality of clustering. Using different combinations of these validity indices as multiple objectives, clustering has been posed as a multi-objective optimization problem to evolve an optimal partitioning of data. In this study, two variants of a multi-objective clustering algorithm are proposed for product-wise clustering of customer complaints. The algorithm employs Multi Objective Particle Swarm Optimization (MOPSO) as the underlying optimization engine. Experiments were carried out on four real-world datasets of customer complaints regarding products and services of four leading commercial banks in India. Experiments were also carried out with the standard K-means and Spherical K-means [3] algorithms to compare the results.

The rest of the paper is organized as follows. First, the motivation behind this work and the contributions made are stated in Sects. 2 and 3 respectively. In Sect. 4, a comprehensive literature survey is presented. A brief description of the techniques employed in this paper is described in Sect. 5. In Sects. 6 and 7, the details of the proposed approach and the experimental setup are described respectively. Section 8 presents the results and discussion and the concluding remarks along with future directions are presented in Sect. 9.

2 Motivation

A tool to analyze customer complaints should ideally help the operational customer relationship management executives to identify the landscape of complaints so that they could derive actionable insights from them rapidly and also prioritize the complaints for a speedy redressal. Manually going through the huge number of complaints, let alone redressing them, is tedious as well as and ineffective.

Motivated by this requirement, a multi-objective clustering framework has been developed to efficiently produce product-wise segmentation of a vast corpus of customer complaints so that the complaints could be shared with the operational customer relationship management (OCRM) department for quick redressal.

3 Contributions

The contribution of this work is manifold.

1. *Firstly*, to the best of our knowledge this is the first study, which employs multi-objective clustering of customer complaints on social media for subsequent sentiment analysis.
2. *Secondly*, the automatic product-wise clustering of the complaints enhances the speed and effectiveness of the grievance redressal systems.
3. *Thirdly*, previous works have employed clustering in sentiment analysis for identification of aspects, sentiment classification or dimension reduction. This is the first study which employs clustering for a full-fledged descriptive and prescriptive analytics of customer complaints data. It falls under descriptive analytics because clustering yields natural homogeneous groupings in the data, while it also falls under prescriptive analytics because the automatic product-wise clustering yields business insights to the OCRM team so that the team can allot resources to redress grievances product wise.

4 Literature Survey

Our proposal is based on social media document clustering with multi-objective optimization. Hence, we divided our literature survey into two sections. First section reviews some existing literature, which performed sentiment analysis using document clustering. Second section reviews techniques related to multi-objective optimization.

4.1 Clustering Based Sentiment Analysis

Although supervised learning techniques have been at the forefront of sentiment analysis research, several notable mentions of the use of unsupervised techniques like clustering can also be found in the literature. Reference [4] was the first to propose a clustering-based approach for sentiment classification at a document level as an alternative to the existing supervised learning and symbolic techniques. They employed the standard K-means algorithm to cluster the TF-IDF weighted document vectors. They used term scores from WordNet to extract sentiment words and showed that the accuracy of classification improves with the selected feature subset.

Clustering has been widely used in sentiment analysis for identification of aspects in data like product reviews or movie reviews. Typically, clustering is employed for identification of aspects in the reviews and then a classifier is trained to perform aspect-level sentiment classification. Reference [5] proposed a method to identify aspects from product reviews by clustering documents using repeated bisection algorithm. They used this information to label the entire dataset and performed multi-class classification using SVM. Reference [6] proposed to group feature synonyms using clustering to improve aspect-based sentiment analysis. Labeled data was automatically generated using sharing words and lexical similarity, and the unlabeled feature expressions were assigned a class by using an EM based semi-supervised learning model. Reference [7] demonstrated that representing documents as bag of nouns yields more meaningful aspects as compared to the bag of words representation.

Evolutionary algorithms based clustering has also been employed in other sentiment analysis tasks. Nature inspired clustering techniques based on Particle Swarm Optimization [8] and Cuckoo Search [9] have been proposed for the sentiment classification task. Recently [10] proposed three, swarm based clustering models based on a hybrid Cuckoo Search and PSO for opinion mining. They compared the performance of these models with other metaheuristic based and traditional clustering algorithms viz., Artificial Bee Colony, Cuckoo Search, K-means and Agglomerative algorithms.

Reference [11] applied swarm based clustering to alleviate the problem of high dimensionality and sparseness in the data prepared for sentiment analysis. Reference [12] employed clustering to reduce the feature size of the word embeddings generated by Word2Vec. With the reduced matrix, they then trained two classifiers with the Support Vector Machine and Logistic Regression algorithms and compared the performance against Doc2Vec and BOW representations.

In addition to term frequency and TF-IDF based representations, documents have also been represented in other forms for the task of clustering. Reference [13] employed Word2Vec representations of emojis to group them into sentiment clusters using agglomerative clustering. They proposed a unique multi-class multi-label classification model augmented with clustering for sentiment analysis of Twitter data. Reference [14] used a neural language model called GloVe to represent the documents. They attempted to improve the clustering based aspect identification task by proposing a weighted context representation model based on semantic relevance along with a flexible constrained K-means algorithm to cluster aspect phrases.

Reference [15] performed a comprehensive study on the performance of clustering methods for sentiment analysis. They studied the impact of the different clustering algorithms, the weighting models (word representations) and other preprocessing techniques like stemming, stop word removal using various benchmark datasets.

4.2 Multi-objective Clustering

Traditional clustering algorithms such as K-means attempt to find the partitioning of data by optimizing a single objective function. These methods are highly sensitive to initialization and are susceptible to getting trapped in the local minima. To alleviate this problem, several researchers proposed evolutionary multi-objective optimization (EMO) techniques which try to find the global optimum of the chosen objective function. However, a single objective function defining a specific property of clustering solution may not be suitable for data with different properties. Optimizing multiple objectives results in robust clusters for datasets with different properties [16].

VIENNA [17] (Voronoi Initialised Evolutionary Nearest-Neighbor Algorithm) marks the first attempt in the direction of multi-objective optimization. This algorithm employed PESA-II as the multi-objective optimization framework. Later, the authors proposed MOCK-am [18], the first multi-objective clustering algorithm for categorical data and MOCK [19] (multi-objective clustering with automatic k-determination), an improved version of VIENNA which allows to automatically determine the optimal number of clusters in the data along with a method to choose the best solution from the Pareto optimal front. Reference [20] proposed MOGA for multi-objective fuzzy clustering. This algorithm employed centroid based encoding. Reference [21] proposed MOES, in which they proposed a variable length encoding of the chromosomes for the first time. The above mentioned works simultaneously optimize two objectives to obtain the optimal partitioning of data. Reference [22] proposed three algorithms viz., MOGA-MCLA, MOGA-CSPA and MOGA-HGPA which employ three objective functions for segmentation of brain MRI scan images. To obtain the final clustering they applied cluster ensemble techniques based on graphs. MOKGA proposed by [23] was the first algorithm to employ four objective functions for clustering.

Over the years, multi-objective clustering has been successfully applied to a variety of tasks like social network analysis [24], identifying co-expressed genes in bioinformatics [25], image segmentation [26], and time-series data analysis [27].

While multi-objective genetic algorithms have been widely used for clustering, other metaheuristics have also been exhaustively explored for the purpose of multi-objective clustering. Some of the algorithms include Genetic Programming [28], Simulated Annealing [29] and Differential Evolution [30].

Particle Swarm Optimization [31] is an important and effective metaheuristic belonging to the class of Swarm Intelligence algorithms. The multi-objective version of PSO, also the prime focus of this work, has been explored for clustering by some works. Reference [32] employed fuzzy clustering with MOPSO as the underlying framework for electric power dispatch, while [33] employed MOPSO for the clustering of hyperspectral images. Reference [34] proposed a hybrid Multi-objective Immunized Particle Swarm Optimization based clustering algorithm for clustering actions of 3D human models.

5 Description of the Techniques

5.1 Multi-objective Optimization

It has been well noted in literature that for solving some problems, simultaneous optimization of multiple objectives is necessary [16]. As regards clustering, it is desirable to obtain a clustering solution in which the intra-cluster distance is minimum and the inter-cluster distance (separation) between the clusters is maximum. Optimizing only one of the two clusters may not return the optimal clusters or partition of the data. The task of multi-objective optimization can be formally stated [35] as

finding a vector of decision variables

$$\bar{x}^* = \left[x_1^*, x_2^*, \ldots x_n^*\right]^T$$

which optimizes a vector function

$$\bar{f}(\bar{x}) = [f_1(\bar{x}), f_2(\bar{x}), \ldots, f_k(\bar{x})]^T$$

while satisfying m inequality constraints

$$g_i(\bar{x}) \geq 0, i = 1, 2, \ldots, m$$

and p equality constraints

$$h_i(\bar{x}) = 0, i = 1, 2, \ldots, p \tag{1}$$

Multi-objective optimization algorithm is run till a stopping criterion is met, which is usually a fixed number of iterations. This yields a set of compromise solutions called Pareto optimal front such that the solutions in the front are non-dominated. That is, no solution is better than any other solution in the set in all of the defined objectives. In other words, one cannot improve an objective function without worsening any other. From the Pareto optimal front, the required solution is either selected by a decision maker or using a statistic like the Gap statistic [19].

5.2 MOPSO

Inspired by the strategies applied by a flock of birds in their search for food, Kennedy and Eberhart [31] proposed the Particle Swarm Optimization algorithm (PSO) which is a powerful metaheuristic belonging to the class of swarm intelligence. Owing to its simple definition and a relatively fewer number of parameters to tune, it is widely used as a global optimization technique. In PSO, each solution is termed as a particle and has a position and velocity vector associated with it. In each iteration, the velocity and position of the particles is guided (updated) by the best position achieved so far by the particle and the best position achieved collectively by the swarm. Formally, the velocity V and position P of particle i at iteration t + 1 are updated as follows.

$$V_{t+1}[i] = W \times V_t[i] + C_1 R_1 \times (PBESTS_t[i] - P_t[i]) + C_2 R_2 \times (GBEST_t - P_t[i]) \tag{2}$$

$$P_{t+1}[i] = P_t[i] + V_{t+1}[i] \tag{3}$$

where, W is the inertia weight, R_1 and R_2 are uniformly distributed random numbers in the range [0, 1]. *PBESTS* is the best position that the particle i has reached and *GBEST* is the global best position. C_1 and C_2 are the cognitive and global components respectively.

Reference [36] proposed a multi-objective version of PSO which has been applied to solve a variety of problems. The basic framework of MOPSO is similar to the standard PSO in terms of the update rules for velocity and positions of the particles. As multiple objectives are optimized simultaneously by MOPSO and there is no global best solution as in standard PSO, at the end of each cycle, a set of non-dominated solutions are generated. Within these new solutions, only those solutions which are non-dominated by the existing solutions are stored in the repository. In MOPSO, the space spanned by the particles in the repository is divided into hypercubes and the global best guide for the velocity update step is chosen randomly from the hypercube selected by roulette wheel selection.

Reference [37] proposed an extension to the MOPSO algorithm proposed in [36] by adopting the crowding distance calculation operator proposed in NSGA-II [38] for the selection of the global best particle during the velocity updation step and also for the selection of the particles to be removed from the repository when it is full. The crowding distance operator calculates the distance of each particle in the repository with its neighbors with respect to each of the objectives. As such, the crowding distance is an estimate of the density around the particle in the objective space. After the calculation of crowding distance for each particle, those particles with highest crowding distance (or in the least density areas) are selected as global best guides so that search is promoted in less crowded areas and diversity of Pareto front is promoted. Also, when the repository is full and a decision has to be taken on the particle to be removed, the particles with least crowding distance (or in the densest areas) are removed to preserve diversity in the population of non-dominated solution. In this work, we adopt the MOPSO-CD algorithm as the multi-objective optimization framework, where CD stands for crowding distance.

5.3 Clustering Heuristics

In this work, we adopt heuristics of two popular clustering algorithms, K-means [2] and Spherical K-means [3] for the two models proposed.

5.3.1 K-Means

K-means is the most popular clustering algorithm and also one of the most popular machine learning algorithms. The simplicity and speed of the algorithm make it the most preferred algorithm for clustering. The algorithm starts by picking K vectors as initial cluster centroids (chosen by different heuristics) and assigning each data vector to the nearest cluster centroid. The distance between the cluster centroid and the data vectors is given by the Euclidean distance metric. For two vectors p and q, the Euclidean distance metric is defined as follows.

$$d(p, q) = d(q, p) = \sqrt{\sum_{i=1}^{n}(q_i - p_i)^2} \tag{4}$$

where, n is the number of vector components in each vector.

After this assignment, the algorithm shifts the centroids to the mean of each of the clusters. This process is repeated till a stopping criterion is met.

K-means works as an optimization technique which tries to minimize the sum of squared distances between the data vectors and their nearest cluster centroids as an objective. It performs a greedy search over the objective space and is susceptible to local minima. Also, since it tries to find only compact and spherical clusters as defined by its objective function, it may not always give optimal clustering. In this work, we consider two different objectives and employ the cluster assignment step of K-means with Euclidean distance metric.

5.3.2 Spherical K-Means

Spherical K-means is a variant of K-means algorithm and has demonstrated superior performance for clustering of documents. The algorithm employs the cosine similarity as a metric to compute the distance of the data vectors with the cluster centroids which are determined in each iteration. The cosine similarity metric is independent of the length of the vectors and computes the similarity based on the angle between them. Consider two vectors P and Q. The cosine similarity between these two vectors can be calculated as follows.

$$\begin{aligned} similarity &= \cos(\theta) \\ &= \frac{P \cdot Q}{\|P\|\|Q\|} \\ &= \frac{\sum_{i=1}^{n} P_i Q_i}{\sqrt{\sum_{i=1}^{n} P_i^2}\sqrt{\sum_{i=1}^{n} Q_i^2}} \end{aligned} \tag{5}$$

where, P_i and Q_i represent the vector components of the vectors P and Q respectively and n is the number of vector components in each vector.

As the computation of similarity is independent of the length of the vectors, it is a suitable metric for determining the similarity between documents.

5.4 Cluster Validity Indices

5.4.1 Intra-cluster Variance

This is the standard objective for the K-means algorithm. It is calculated as follows.

$$E = \frac{1}{N} \sum_x \left\| x - \mu_{k(x)} \right\|^2 \tag{6}$$

where, N is the number of samples, x is a sample and $\mu_{k(x)}$ is the centroid of a cluster to which x belongs to.

This index is a measure of how similar or homogeneous the objects within a cluster are and hence needs to be minimized as an objective. This index favors compact clusters.

5.4.2 Calinski-Harabasz Index

This internal cluster validity measure introduced in [39] is defined as a ratio of the between cluster variance to the within cluster variance of the examined clustering solution. It is calculated as follows.

$$CH = \frac{\sum_{k=1}^{K} n_k \|z_k - z\|^2}{K - 1} \Big/ \frac{\sum_{k=1}^{k} \sum_{i=1}^{n_k} \|x_i - z_k\|^2}{n - K} \tag{7}$$

where, K is the number of clusters, z is the centroid of the data, z_k is the kth cluster centroid, n is the number of data samples in the data, x_i is the ith sample.

This index simultaneously measures two important aspects of a clustering solution—inter-cluster separation and intra-cluster similarity. For a good clustering solution, inter-cluster variance should be maximum and intra-cluster variance should be minimum. Hence, this index needs to be maximized as an objective.

5.4.3 Silhouette Index

The Silhouette value [40] is a measure of how relevant an object is to its own cluster (consistency) compared to other clusters (separation). The value of Silhouette index varies from -1 to $+1$. High value of Silhouette index indicates that clustering is good and all points are similar to their own cluster than to their neighbors where as low value of Silhouette index indicates the opposite. Silhouette index can be computed by using many distance metrics such as Euclidean distance, cosine similarity etc., depending on the application. Silhouette Index is calculated as follows.

$$SIL = \frac{1}{K} \sum_{k=1}^{K} SIL(C_k)$$

$$SIL(C_k) = \frac{1}{n_k} \sum_{x \in C_k} SIL(x)$$

$$SIL(x) = \frac{b(x) - a(x)}{\max(a(x), b(x))} \tag{8}$$

where, $a(x)$ is a measure of how well x is assigned to its cluster (the smaller the value, the better the assignment). Next, the mean dissimilarity of point x to a cluster c is defined as the mean of the distance from x to all points in c. Let $b(x)$ be the lowest average distance from x to all points in another cluster, of which the point x is not a member. The cluster with this lowest average difference is said to be the neighboring cluster of x because it is the next best matching cluster for point x.

6 Proposed Approach

An outline of the proposed approach is described in this section and a schematic of the approach is depicted in Fig. 1.

6.1 Text Preprocessing

This step was performed using the Natural Language Toolkit (NLTK) in the Python programming language. We followed standard text mining procedures like noise removal, tokenization and lemmatization on the whole corpus. In this study, we rely on the occurrence and significance of terms to capture the severity of complaints. Hence, stop words were removed as they contribute to noise in this scenario.

6.2 Feature Extraction and Representation

In this study, we considered the TF-IDF feature representation to represent the complaint documents. This representation is a normalized version of the TF based DTM. Each of the term frequencies calculated in TF based DTM is weighted by the inverse of the occurrence of the term in the entire corpus (IDF). Formally, IDF value of a term t, appearing in a document D, is calculated using the following equation.

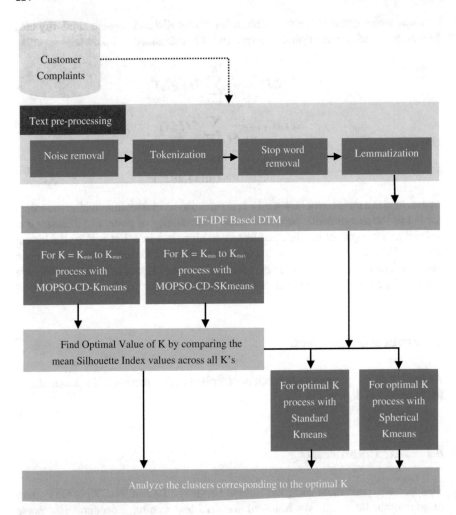

Fig. 1 A schematic of the proposed approach

$$IDF(t, D) = \log \frac{N}{|\{d \in D : t \in D\}|} \qquad (9)$$

6.3 Details of the Optimization Algorithm

The proposed algorithm is described in detail in this section and an outline of it is depicted in Fig. 2.

Input: popsize = *P*, maximum number of generations = *maxgen*, size of the repository = *repsize*, objective functions = *obj₁*, *obj₂*, probability of mutation = *mut*, cognitive component = C_1 and global component = C_2

Output: Clustering solution

Initialisation

1. Initialize the positions and velocities of all particles in *P*.
2. Evaluate the fitness of each particle across the two objectives using equations (6) and (7) and store as *personal bests*.
3. Store the non-dominated solutions in repository *R*.

LOOP Process

4. **for** *t* = *0* to *maxgen* **do**
5.   Update the velocities and positions of particles using (2) and (3) respectively. For each particle, the global best is chosen randomly from the top 10% of *R*, after sorting the solutions in descending order of their crowding distances.
6.   Check the particles for infeasibility and reintegrate into feasible space if necessary.
7.   Mutate the particles with probability *mut*.
8.   Calculate a membership matrix containing memberships of each of the data points with each of the particles using the clustering heuristics. For MOPSO-CD-Kmeans, memberships are calculated using equation (4) and for MOPSO-CD-SKmeans, the memberships are calculated using equation (5). Since there are *P* particles, *P* different partitions of the data are obtained.
9.   Using these membership matrices, evaluate all the particles in *P* across both objectives using equations (6) and (7).
10.   Update *R* by replacing dominated solutions with newly non-dominated solutions. If length of *R* is *repsize*, compute the crowding distance of each particle, sort them in descending order and replace the particles from the bottom 10% of *R* with new non-dominated solutions.
11. **end for**
12. For each non-dominated solutions present in *R*, calculate the Silhouette Index.
13. **return** the solution with the maximum Silhouette Index.

Fig. 2 Outline of the proposed algorithm

6.3.1   Problem Formulation

In this study, the clustering problem is formulated as a bi-objective optimization problem. Intra-cluster variance is considered as the first objective function $f_1(x)$

$$\text{minimize } f_1(x) = Intra\text{-}cluster\ variance \tag{10a}$$

while, Calinski-Harabasz index is considered as the second objective function $f_2(x)$

$$\text{maximize } f_2(x) = Calinski\text{-}Harabasz\ index \tag{10b}$$

The vectors representing the cluster centroids of a single clustering solution are defined as the decision variables. The optimization problem can then be defined as follows. Find a vector of K × d decision variables which optimizes the vector function

$$\bar{f}(\bar{x}) = [f_1(\bar{x});\ \ f_2(\bar{x})]^T \tag{10c}$$

These two objectives, Intra-cluster variance and Calinski-Harabasz index are simultaneously optimized by MOPSO-CD to evaluate the fitness of the particles. As mentioned earlier, intra-cluster variance favors more compact clusters while the Calinski-Harabasz index considers both, the separation between the clusters and similarity within the clusters. Hence, with the help of these two objectives MOPSO-CD drives the swarm towards the solution space containing compact and well-separated clusters.

6.3.2   Particle Representation and Initialization

As mentioned above, each particle is a string of real numbers which denotes the set of cluster centroids that represent a single partitioning of the data. As such, each particle represents one clustering solution. If a particular string encodes the centers of K clusters, in a d dimensional space, the length of the corresponding particle will be K × d.

For example, in a three-dimensional space, the string1.2 2.5 3.2 7.4 2.7 1.6 3.9 5.2 0.8 1.5 3.2 2.6> represents a partition of four clusters with cluster centers <1.2 2.5 3.2>, <7.4 2.7 1.6>, <3.9 5.2 0.8>, <1.5 3.2 2.6>.

6.3.3   Computation of Membership

Data points are assigned to their nearest cluster centers to determine the members of the clusters. For each data point $\bar{x}_j, j = 1, 2, \ldots n$, the membership is determined in the following way. For each particle, cycle through the cluster centers it encodes and find the cluster center nearest to $\bar{x}_j$ using the minimum value criterion: $k = Argmin_{i=1...K}d(\bar{x}_j, \bar{c}_i)$. Here, d is the distance metric used.

For MOPSO-CD-Kmeans, the distance metric is Euclidean distance criterion as conventionally employed in the K-means algorithm. For MOPSO-CD-SKmeans, cosine similarity is the distance metric employed according to the standard definition of the algorithm. The memberships of the data points are calculated with respect to each particle, as each particle represents a set of cluster centers. Hence, each particle has a corresponding membership matrix. These memberships are used in the next step for evaluating the fitness of the particles.

6.3.4 Velocity and Position Update

Velocities and positions of the particles are updated as per the standard PSO equations. However, since MOPSO-CD derives a number of non-dominated solutions in each cycle, there is no global winner. The global best particle is selected for each particle as per the following procedure. All the non-dominated solutions are sorted according to their crowding distance and the top 10% of the solutions are determined (favoring the solutions with larger crowding distance). Then, for each particle whose velocity is to be updated, a solution is randomly picked from the top 10%. The selected solution serves as the global best guide for the corresponding particle. Once the velocities and positions are updated, the particles are then checked for feasibility, to ensure that they lie within the search space. If any particle goes beyond the upper or lower bounds of any of the dimensions, then it is restored by setting the value of that dimension to the corresponding boundary and is made to search in the opposite direction by multiplying its velocity by -1.

6.3.5 Updating the Repository

The notions of non-domination and crowding distance are used to populate the repository. After each round of fitness evaluation, the non-dominated particles are inserted into the repository till its capacity is exhausted. At this point, when a new particle dominates any of the existing particles in the repository, the existing particle is replaced with the new particle. If the new particle is non-dominated with respect to particles in the repository, the particles are sorted based on crowding distance and the particles with the least crowding distance are replaced. This ensures that search is directed towards sparse regions of the search space and a good spread of non-dominated solutions is maintained.

6.3.6 Mutation

The mutation operator has been adopted to enhance the exploratory ability of the algorithm and to prevent possible premature convergence. In the initial iterations the coverage of the mutation operator is larger to give the algorithm more exploratory power and decreased gradually over time for a more exploitative search.

6.4 Selecting the Best Solution and Optimal Number of Clusters

The output of any EMO algorithm is a set of optimal compromise solutions (Pareto front) in which no solution is a clear winner. In the algorithm proposed in this study, each solution represents a particular partitioning of data and is non-dominated with respect to other solutions in terms of intra-cluster variance and Calinski-Harabasz index. However, since the user would require only one solution, the Silhouette index is employed to select the best solution from the Pareto optimal front. For each non-dominated solution in the final repository, the Silhouette index is computed and the solution with the maximum value for Silhouette index is reported as the best solution. Also, in the absence of domain knowledge, it is not possible to guess the number of clusters in the data. Hence, a heuristic is required to predict the best guess for the value of K, i.e., the number of clusters needed to be generated. The optimal value for the number of clusters for the data is derived as follows. A range of values K_{min} to K_{max} is decided and then the algorithm is run till the stopping criterion is reached for each value of K. To restrict the influence of random effects, the experiments were repeated 30 times for each value of K. This is a standard procedure for experiments involving EMO. Hence, for each value of K, the mean of Silhouette index values of the best particles of all 30 trials is calculated. The value of K for which the Silhouette index is the highest is considered as the closest guess for the optimal value of K for a given dataset.

7 Experimental Setup

The effectiveness of the proposed algorithm is demonstrated using four different datasets. In the following section, we elaborate on the preprocessing workflow adopted to prepare the data for the experiments.

7.1 Datasets

7.1.1 Dataset Details

We experimented with 4 bank based datasets which were previously employed for automatic complaints summarization [41]. The authors of [41] crawled the web

Table 1 Details of datasets

Web resource	Company name	No. of complaints [41]	No. of complaints considered after preprocessing
www.complaintboard.in	Axis Bank	749	513
	HDFC Bank	1108	627
	ICICI Bank	758	440
	SBI Bank	1123	676

and collected 749, 1108, 758, and 1123 consumer complaints and bank executive responses from www.complaintboard.in on four Indian banks namely Axis Bank, HDFC Bank, ICICI Bank, and SBI respectively. We performed the necessary preprocessing, dropped bank executives' responses, and annotation. These preprocessing steps yielded 513, 637, 440 and 676 consumer complaints for AXIS, HDFC, ICICI, and SBI dataset respectively. The first three are leading Indian private sector banks, whereas SBI is the biggest public sector banks in assets. These datasets were crawled during October 2014 to December 2014. So, the datasets include complaints posted before December 2014 on given complaints forums. Table 1 presents details of the datasets.

7.1.2 Filtering

From the collected complaints, it was essential to filter out some documents to make the dataset homogeneously represent consumer complaints. Following types of documents were removed:

- Responses from bank executives
- Uninformative documents which do not reflect the concern of the customer adequately
- Customer requests for information or service
- Duplicates

7.1.3 Data Cleansing

The medium of complaint submission on the forum is a text box. Therefore, the complaints lack a proper structure and are noise-ridden. We observed and eliminated the following types of noise in the collected complaints:

- Threads of email communication with customer support dumped into the complaints
- Non-English text
- Email addresses and phone numbers

- Misspelled words and abbreviations
- Words of length less than 3

7.2 User-Defined Parameters

The parameters chosen for MOPSO-CD are identical for both the variants of the algorithm. This is to facilitate comparison of performance of different feature representations and the two clustering heuristics. The population size and repository size used are 50 and 30 respectively. Other parameters of MOPSO-CD are set to the values recommended in the base paper. The mutation probability is set to 0.5, inertia is set to 0.4 and c_1 and c_2 are both set to 1.0. The values of K_{min} and K_{max} are set to 2 and 10 respectively. For K-means and Spherical K-means, the value of K was set equal to the optimal K value determined by the better performer among MOPSO-CD-Kmeans and MOPSOCD-SKmeans. Other parameters were set to default values.

All the experiments have been performed on an Intel® Xeon(R) CPU E5-2640 v4, with 2.4 GHz, 32 GB RAM in an Ubuntu 16.04 environment.

8 Results and Discussion

In the following section, we empirically evaluate the performance of MOPSOCD-Kmeans and MOPSO-CD-SKmeans. To facilitate comparison with baseline algorithms, clustering was performed on all four datasets with the standard single objective K-means algorithm and the Spherical K-means algorithm using the implementations available in open source Python libraries.

The results obtained by MOPSO-CD-SKmeans for all datasets are reported in Table 2. By comparing the mean Silhouette index values across different values of K, it was deduced that the optimal number of clusters for Axis, HDFC, ICICI and SBI datasets are 10, 10, 9 and 10 respectively. Among the non-dominated solutions generated by the algorithm for the winning K, the solution with the highest Silhouette index value was chosen as the best solution.

Cluster analysis of best solution determined in this manner, is then performed. A detailed account of the analysis bank-wise is as follows.

- A summary of clusters obtained for Axis dataset are presented in Table 3. The clusters represented complaints regarding **credit and debit cards, loans, account** related complaints like **uninformed debits, customer service, cheque books, harassment by recovery agents** and **internet banking** respectively. One cluster had rare complaints mapped to it. For example the entire corpus contained single or very few complaints regarding some products like **mutual funds** or regarding **funds transfer** or **account transfer**. This cluster had complaints related to such rare products mapped to it and did not seem to follow a pattern. One cluster comprised

Table 2 Mean Silhouette index values for each K obtained by MOPSO-CD-SKMeans

Dataset	Optimal K	2	3	4	5	6	7	8	9	10
Axis	10	0.025	0.035	0.040	0.043	0.046	0.048	0.055	0.056	**0.057**
HDFC	10	0.022	0.022	0.022	0.027	0.029	0.034	0.036	0.039	**0.043**
ICICI	9	0.023	0.028	0.031	0.036	0.043	0.047	0.052	**0.057**	0.055
SBI	10	0.019	0.020	0.023	0.027	0.032	0.033	0.035	0.038	**0.042**

Table 3 Summary of clusters for Axis dataset

Cluster ID	Number of complaints	Category
1	20	Login issues
2	81	Credit card and debit card
3	105	Loans
4	126	Deductions, uninformed charges
5	16	Miscellaneous
6	50	Customer service
7	48	Cheques
8	5	Harassment
9	28	Internet banking
10	34	Mutual funds, fund transfer, account transfer

of *login issues* which several customers were facing. The fact that the algorithm was able to separate it from the cluster with *internet banking complaints* makes it a very interesting result and demonstrates the strength of the algorithm.

- A summary of clusters obtained for HDFC dataset are presented in Table 4. The clusters represented complaints regarding *credit cards*, *customer service*, *defaulting issues*, *loans* respectively. In addition to these clusters, two clusters comprised of complaints about *customer service* and three clusters comprised of complaints about *uninformed charges and deductions* from bank accounts. One cluster comprised of rare complaints like *request for ATM card, closure of bank accounts*, etc. The last cluster comprised of complaints exclusively regarding *HDFC securities*.
- A summary of clusters obtained for ICICI dataset are presented in Table 5. One cluster comprised of complaints related to *loans*, two major clusters comprised of *credit card* related complaints with the bigger cluster among them having few complaints related to other products related to *credit cards* in terms of nature of the

Table 4 Summary of clusters for HDFC dataset

Cluster ID	Number of complaints	Category
1	16	Deductions
2	103	Credit cards
3	21	Customer service
4	76	Settlement, recovery
5	75	Deductions
6	32	Customer service
7	81	Fraudulent transactions, miscellaneous complaints
8	77	Loan
9	3	HDFC sec
10	153	Deductions

Table 5 Summary of clusters for ICICI dataset

Cluster ID	Number of complaints	Category
1	15	Miscellaneous (CC, loan, infrequent complaints)
2	31	iMobile app (specifically recharge failure)
3	156	Mostly credit card and few miscellaneous
4	56	Credit card
5	12	Miscellaneous
6	119	ATM/deductions/miscellaneous
7	5	Miscellaneous
8	34	Deduction
9	12	Loan

complaints. One cluster comprised of complaints related to **uninformed charges** debited from **bank accounts**. Similar complaints were also present in another cluster which also comprised of complaints regarding **ATM**. This heterogeneity in the cluster could be attributed to the similarity in the nature of the complaints and hence, usage of similar words in the documents. For example, the words debited and return and account appear in complaints regarding **unsuccessful ATM transaction** accompanied by debit of money from the account and **uninformed charges** debited from **bank accounts**.

- There were minor clusters comprising of 15, 12 and 9 complaints each which represented infrequent product classes like **tax saving bond, provisional statement** not provided and a very few complaints from the majority product classes like **credit cards** or **loans**. One interesting cluster comprised of complaints referring exclusively to ICICI's **mobile banking** app called **iMobile**, again demonstrating the strength of the algorithm in isolating specific issues.

- A summary of clusters obtained for SBI dataset are presented in Table 6. The clusters comprised of complaints regarding **credit** and **debit cards, mobile banking, loans, failed transactions, ATM** respectively. One major cluster of size 196 exclusively comprised of **credit card** related complaints. One cluster comprised of complaints centered on "payments" in different contexts but did not represent a specific product class. Two other clusters also showed poor homogeneity with respect to complaints about product class. One cluster comprising exclusively of complaints related to a product called **Credit card Protection Plan (CPP)** promoted by SBI.

The results obtained by MOPSO-CD-Kmeans on all datasets are reported in Table 7. The algorithm predicted 9, 10, 5, and 8 as the optimum number of clusters for Axis, HDFC, ICICI and SBI datasets respectively. Among the non-dominated solutions for the optimal K for each dataset, the solution with the highest Silhouette index value was chosen as the best solution. Cluster analysis was then performed on the best solution thus determined for each dataset. It was noticed in each case that the

Table 6 Summary of clusters for SBI dataset

Cluster ID	Number of complaints	Category
1	48	Centered around the word "payment"
2	78	Mobile banking
3	85	Cards
4	13	Credit cards (CPP charges)
5	22	Loan
6	44	Failed transactions
7	87	ATM
8	22	Miscellaneous
9	196	Credit cards
10	79	ATM/pension/infrequent complaints

algorithm assigned majority of the complaints to a single cluster and the remaining clusters had very sparse members. The single major cluster had complaints of all categories mapped to it and a product-wise distribution of complaints in the clusters was missing. The cluster analysis had yielded an impression of arbitrary clustering by the algorithm.

Clustering was then performed with the standard K-means algorithm as a baseline to establish the efficacy of MOPSO-CD-SKmeans. The optimal value of K for standard K-means was set equal to the optimal K values obtained by MOPSO-CD-SKmeans, as reported in Table 2. The clusters obtained depict that K-means failed to achieve product-wise segmentation noticed in the case of MOPSO-CD-SKmeans. K-means assigned majority of the complaints belonging to different product clusters to 3–4 clusters and the remaining clusters comprised of very few complaints.

Finally, clustering was performed with the standard Spherical K-means algorithm to compare its performance against the proposed multi-objective version, MOPSO-CD-SKmeans. The optimal value of K for standard Spherical K-means was set equal to the optimal K values obtained by MOPSO-CD-SKmeans, as reported in Table 2. On analysis of the clusters generated by Spherical K-means, it was noticed in case of each dataset that some of the clusters exhibited homogeneity in terms of the product which the complaints refer to. Unlike the standard K-means or MOPSO-CD-Kmeans, the clustering was not arbitrary and demonstrated product-wise clusters. However, it was also noticed that the segmentation worked well for documents containing some strong patterns. For example, in case of Axis dataset, one of the clusters comprised of 6 complaints related to *login issues*. Two documents from this cluster are shown below:

Since yesterday when I am **trying to login in my AXIS BANK** net banking **I am getting the error like–** . Please check this and let me know what's the issue. Thanks, Iti Mishra.

Today I **tried to login in my AXIS BANK** for transaction, **I am getting the error like** – . Please solve the problem as soon as possible.

Table 7 Mean Silhouette index values for each K obtained by MOPSO-CD-KMeans

Dataset	Optimal K	2	3	4	5	6	7	8	9	10
Axis	9	0.475	0.509	0.566	0.562	0.568	0.592	0.604	**0.653**	0.647
HDFC	10	0.456	0.511	0.554	0.525	0.595	0.612	0.569	0.597	**0.618**
ICICI	5	0.497	0.568	0.578	**0.659**	0.616	0.587	0.587	0.639	0.612
SBI	8	0.458	0.519	0.521	0.585	0.561	0.615	**0.632**	0.581	0.598

For complaints in which such a pattern was missing, the standard Spherical K-means did not perform well and the overall distribution of the complaints among the clusters was not comparable to MOPSO-CD-SKmeans. For each of the datasets there is a major cluster which contains a mixture of complaints related to various products. In case of Axis dataset this cluster comprises 37.23% of the total complaints, in case of HDFC dataset, there are two clusters which comprises 52.03% of the total complaints, in case of ICICI dataset, there is a major cluster which comprises 54.09% of the total complaints and in case of SBI dataset, there is a major cluster which comprises 45.25% of the total complaints. This indicates that the standard Spherical K-means algorithm failed to properly segment, a large portion of documents in each of the datasets.

In summary, it could be observed that the MOPSO-CD-SKmeans algorithm is successful in achieving product-wise clustering to a large extent. Some specific issues like ***uninformed CPPcharges*** are remarkably isolated by the algorithm. When such a cluster is identified, management could take a decision to inform all the customers who could face this issue so that customer grievance is avoided. This type of isolation of trending issues is very difficult to achieve manually or by using traditional analytics tools like spreadsheet programs or SQL queries. There are instances of heterogeneous clusters generated for ICICI and SBI datasets. This could be attributed to a combination of factors.

- First and foremost, there is an overlap in the nature of complaints. For example, both, complaints by credit card and loan customers facing harassment by recovery agents have a large overlap in the terms mentioned in the documents. Since the feature representation adopted here highlights term occurrence, this does not allow the algorithm to discriminate the complaints.
- Secondly, some of the complaints are infrequent. Special methods need to be developed for more powerful discrimination of such complaints.
- Thirdly, although exhaustive data preprocessing was performed on the data, there remains further scope in terms of data cleansing and feature selection.

9   Conclusions and Future Directions

In this paper, two variants of a novel multi-objective clustering algorithm are proposed with applications to CRM in the banking industry. In both the algorithms, MOPSO-CD is employed as the multi-objective optimization engine to cluster the data while simultaneously optimizing two conflicting clustering objectives viz., intra-cluster variance and the Calinski-Harabasz Index. The first algorithm employs the heuristics of K-means with Euclidean distance as the distance metric while the second algorithm employs the heuristics of Spherical K-means which has cosine similarity at its core. Experiments were performed on TF-IDF based DTM of four bank based complaints datasets with both the variants. The results indicate that the variant which employs Spherical K-means can be identified as the best model. The clusters generated by this variant depict uniformity in terms of the product referred to, in the complaints. Comparison studies were also performed with standard K-means and Spherical K-means algorithms. It was noticed that the standard K-means failed to achieve product-wise segmentation. While the standard Spherical K-means was able to generate product-wise clusters, the segmentation obtained by MOPSO-CD-SKmeans demonstrated superior segmentation.

There are several pockets of improvement in this work. Firstly, TF-IDF values of the bag of words extracted from the complaints corpus were considered as feature representations for the documents. In future works, bigram or trigrams could be replaced with bag of words. Secondly, feature representations generated with advanced neural language models like Word2Vec or Doc2Vec could be explored with the proposed models to see if they yield better results. Thirdly, severity based segmentation of complaints within different product groups could be made possible by fusing lexicon based techniques with machine learning techniques. On a broader scale, these systems could be fused with modern NLP based systems like chatbots for a multifold improvement in grievance redressal systems. This study opens up new avenues of research in sentiment analysis of customer complaints for analytical CRM. The work carried out could be considered as a precursor for rich research in the direction of grievance redressal and setting up their priorities, using machine learning techniques, in the banking industry.

References

1. Ravi, K., and V. Ravi. 2015. A survey on opinion mining and sentiment analysis: Tasks, approaches and applications. *Knowledge-Based Syst* 89: 14–46. https://doi.org/10.1016/J.KNOSYS.2015.06.015.
2. Macqueen, J. 1967. Some methods for classification and analysis of multivariate observations. In *5-th Berkeley symposium on mathematical statistics and probability*, 281–297.
3. Dhillon, I.S., and D.S. Modha. 2001. Concept decompositions for large sparse text data using clustering. *Machine Learning* 42: 143–175. https://doi.org/10.1023/A:1007612920971.
4. Li, G., and F. Liu. 2010. A clustering-based approach on sentiment analysis. In *2010 IEEE international conference on intelligent systems and knowledge engineering*, 331–337. IEEE.
5. Hadano, M., K. Shimada, and T. Endo. 2011. Aspect identification of sentiment sentences using a clustering algorithm. *Procedia-Social and Behavioral Sciences* 27: 22–31. https://doi.org/10.1016/J.SBSPRO.2011.10.579.
6. Zhai, Z., B. Liu, H. Xu, and P. Jia. 2011. Clustering product features for opinion mining. In *Proceedings of the fourth ACM international conference on web search and data mining-WSDM '11*, 347. New York, USA: ACM Press.
7. Farhadloo, M., and E. Rolland. 2013. Multi-class sentiment analysis with clustering and score representation. In *2013 IEEE 13th international conference on data mining workshops*, 904–912. IEEE.
8. Souza, E., A.L.I. Oliveira, G. Oliveira, et al. 2016. An unsupervised particle swarm optimization approach for opinion clustering. In 2016 5th Brazilian conference on intelligent systems (BRACIS), 307–312. IEEE.
9. Chandra Pandey, A., D. Singh Rajpoot, and M. Saraswat. 2017. Twitter sentiment analysis using hybrid cuckoo search method. *Information Processing and Management* 53: 764–779. https://doi.org/10.1016/J.IPM.2017.02.004.

10. Souza, E., D. Santos, G. Oliveira, et al. 2018. Swarm optimization clustering methods for opinion mining. *Natural Computing*, 1–29. https://doi.org/10.1007/s11047-018-9681-2.

11. Fong, S., E. Gao, and R. Wong. 2015. Optimized swarm search-based feature selection for text mining in sentiment analysis. In 2015 IEEE international conference on data mining workshop (ICDMW), 1153–1162. IEEE.

12. Alshari, E.M., A. Azman, S. Doraisamy, et al. 2017. Improvement of sentiment analysis based on clustering of Word2Vec features. In 2017 28th international workshop on database and expert systems applications (DEXA), 123–126. IEEE.

13. Mayank, D., K. Padmanabhan, and K. Pal. 2016. Multi-sentiment modeling with scalable systematic labeled data generation via Word2Vec clustering. In 2016 IEEE 16th international conference on data mining workshops (ICDMW), 952–959. IEEE.

14. Xiong, S., and D. Ji. 2016. Exploiting flexible-constrained K-means clustering with word embedding for aspect-phrase grouping. *Information Sciences (Ny)* 367–368: 689–699. https://doi.org/10.1016/J.INS.2016.07.002.

15. Ma, B., H. Yuan, and Y. Wu. 2017. Exploring performance of clustering methods on document sentiment analysis. *Journal of Information Science* 43: 54–74. https://doi.org/10.1177/0165551515617374.

16. Mukhopadhyay, A., U. Maulik, and S. Bandyopadhyay. 2015. A survey of multiobjective evolutionary clustering. *ACM Computing Surveys* 47: 1–46. https://doi.org/10.1145/2742642.

17. Handl, J., and J. Knowles. 2004. *Evolutionary multiobjective clustering*, vol. 3242. Lecture Notes Computer Science, 1081–1091. https://doi.org/10.1007/978-3-540-30217-9_109.

18. Handl, J., and J. Knowles. 2005. Multiobjective clustering around medoids. In *2005 IEEE congress on evolutionary computation*, 632–639. IEEE.

19. Handl, J., and J. Knowles. 2007. An evolutionary approach to multiobjective clustering. *IEEE Transactions on Evolutionary Computation* 11: 56–76. https://doi.org/10.1109/TEVC.2006.877146.

20. Bandyopadhyay, S., U. Maulik, and A. Mukhopadhyay. 2007. Multiobjective genetic clustering for pixel classification in remote sensing imagery. *IEEE Transactions on Geoscience and Remote Sensing* 45: 1506–1511. https://doi.org/10.1109/TGRS.2007.892604.

21. Won, J.-M., S. Ullah, and F. Karray. 2008. Data clustering using multi-objective hybrid evolutionary algorithm. In *2008 international conference on control, automation and systems*, 2298–2303. IEEE.

22. Mukhopadhyay, A., U. Maulik, and S. Bandyopadhyay. 2009. Multiobjective genetic clustering with ensemble among pareto front solutions: application to MRI brain image segmentation. In *2009 seventh international conference on advances in pattern recognition*, 236–239. IEEE.

23. Özyer, T., M. Zhang, and R. Alhajj. 2011. Integrating multi-objective genetic algorithm based clustering and data partitioning for skyline computation. *Applied Intelligence* 35: 110–122. https://doi.org/10.1007/s10489-009-0206-7.

24. Folino, F., and C. Pizzuti. 2010. A multiobjective and evolutionary clustering method for dynamic networks. In *2010 international conference on advances in social networks analysis and mining*, 256–263. IEEE.

25. Mukhopadhyay, A., U. Maulik, and S. Bandyopadhyay. 2013. An interactive approach to multi-objective clustering of gene expression patterns. *IEEE Transactions on Biomedical Engineering* 60: 35–41. https://doi.org/10.1109/TBME.2012.2220765.

26. Shirakawa, S., and T. Nagao. 2009. Evolutionary image segmentation based on multiobjective clustering. In *2009 IEEE congress on evolutionary computation*, 2466–2473. IEEE.

27. Bandyopadhyay, S., U. Maulik, and R. Baragona. 2010. Clustering multivariate time series by genetic multiobjective optimization. *Metron* 68: 161–183. https://doi.org/10.1007/BF03263533.

28. Coelho, A.L.V., E. Fernandes, and K. Faceli. 2010. Inducing multi-objective clustering ensembles with genetic programming. *Neurocomputing* 74: 494–498. https://doi.org/10.1016/J.NEUCOM.2010.09.014.

29. Saha, S., and S. Bandyopadhyay. 2008. A new multiobjective simulated annealing based clustering technique using stability and symmetry. In *2008 19th international conference on pattern recognition*, 1–4. IEEE.

30. Xue, F., A.C. Sanderson, and R.J. Graves. 2005. Multi-objective differential evolution-algorithm, convergence analysis, and applications. In *2005 IEEE congress on evolutionary computation*, 743–750. IEEE.

31. Kennedy, J., and R. Eberhart (1995) Particle swarm optimization. In *IEEE international conference on neural networks*, vol. 4, 1942–1948.

32. Agrawal, S., B.K. Panigrahi, and M.K. Tiwari. 2008. multiobjective particle swarm algorithm with fuzzy clustering for electrical power dispatch. *IEEE Transactions on Evolutionary Computation* 12: 529–541. https://doi.org/10.1109/TEVC.2007.913121.

33. Paoli, A., F. Melgani, and E. Pasolli. 2009. Clustering of hyperspectral images based on multi-objective particle swarm optimization. *IEEE Transactions on Geoscience and Remote Sensing* 47: 4175–4188. https://doi.org/10.1109/TGRS.2009.2023666.

34. Nanda, S.J., and G. Panda. 2013. Automatic clustering algorithm based on multi-objective Immunized PSO to classify actions of 3D human models. *Engineering Applications of Artificial Intelligence* 26: 1429–1441. https://doi.org/10.1016/j.engappai.2012.11.008.

35. Deb, K., and D. Kalyanmoy. 2001. *Multi-objective optimization using evolutionary algorithms*. Wiley.

36. Coello Coello, C.A., and M.S. Lechuga. 2002. MOPSO: a proposal for multiple objective particle swarm optimization. In *Proceedings of the 2002 congress on evolutionary computation, CEC '02*, 1051–1056. IEEE. (Cat. No.02TH8600).

37. Raquel, C.R., and P.C. Naval. 2005. An effective use of crowding distance in multiobjective particle swarm optimization. In *Proceedings of the 2005 conference on genetic and evolutionary computation-GECCO '05*, 257. New York, USA: ACM Press.

38. Deb, K., A. Pratap, S. Agarwal, and T. Meyarivan. 2002. A fast and elitist multiobjective genetic algorithm: NSGA-II. *IEEE Transactions on Evolutionary Computation* 6: 182–197. https://doi.org/10.1109/4235.996017.

39. Caliński, T., and J. Harabasz. 1974. A dendrite method for cluster analysis. *Communications in Statistics-Theory and Methods* 3, 1–27. https://doi.org/10.1080/03610927408827101.

40. Rousseeuw, P.J. 1987. Silhouettes: a graphical aid to the interpretation and validation of cluster analysis. *Journal of Computational and Applied Mathematics* 20: 53–65. https://doi.org/10.1016/0377-0427(87)90125-7.

41. Ravi, K., V. Ravi, and P.S.R.K. Prasad. 2017. Fuzzy formal concept analysis based opinion mining for CRM in financial services. *Applied Soft Computing* 60: 786–807. https://doi.org/10.1016/J.ASOC.2017.05.028.

Benchmarking Gene Selection Techniques for Prediction of Distinct Carcinoma from Gene Expression Data: A Computational Study

Lokeswari Venkataramana, Shomona Gracia Jacob, Saraswathi Shanmuganathan and Venkata Vara Prasad Dattuluri

Abstract Gene Expression (GE) data have been attracting researchers since ages by virtue of the essential genetic information they carry, that plays a pivotal role in both causing and curing terminal ailments. GE data are generated using DNA microarrays. These gene expression data are obtained in measurements of thousands of genes with relatively very few samples. The main challenge in analyzing microarray gene data is not only in finding differentially expressed genes, but also in applying computational methods to the increasing size of microarray gene expression data. This review will focus on gene selection approaches for simultaneous exploratory analysis of multiple cancer datasets. The authors provide a brief review of several gene selection algorithms and the principle behind selecting a suitable gene selection algorithm for extracting predictive genes for cancer prediction. The performance has been evaluated using 10-fold Average Split accuracy method. As microarray gene data is growing massively in volume, the computational methods need to be scalable to explore and process such massive datasets. Moreover, it consumes more time, labour and cost when this investigation is done in serial (sequential) manner. This motivated the authors to propose parallelized gene selection and classification approach for selecting optimal genes and categorizing the cancer subtypes. The authors also present the hurdles faced in adopting parallelized computational methods for microarray gene data while substantiating the need for parallel techniques by evaluating their performance with previously reported research in this sphere of study.

L. Venkataramana (✉) · S. G. Jacob · Saraswathi Shanmuganathan ·
Venkata Vara Prasad Dattuluri
Department of Computer Science and Engineering, Sri Sivasubramaniya Nadar
College of Engineering, Kalavakkam, Chennai 603110, India
e-mail: lokeswaricts@gmail.com

S. G. Jacob
e-mail: graciarun@gmail.com

Saraswathi Shanmuganathan
e-mail: sarasuthan@gmail.com

Venkata Vara Prasad Dattuluri
e-mail: dvvprasad@ssn.edu.in

© Springer Nature Switzerland AG 2020
M. Rout et al. (eds.), *Nature Inspired Computing for Data Science*,
Studies in Computational Intelligence 871,
https://doi.org/10.1007/978-3-030-33820-6_10

Keywords Microarray gene expression data · Gene selection · Parallelized computational methods · Oncogenes

1 Introduction

Every organism on earth has cells except viruses. Each cell has a nucleus. Within the nucleus there is DNA (DeoxyriboNucleic Acid), which carries instructions for making future organisms. DNA is made up of coding and non-coding sequences. Coding sequences are called "Genes" [1–3]. These coding sequences consist of information that forms the proteins. Genes make proteins in two steps. Initially, the DNA is converted to messenger RNA or mRNA (RiboNucleic Acid) through transcription process. The mRNA is further translated into proteins which controls the major functional elements of any organism. The data generated from experiments on DNA, RNA, and protein microarrays could be interpreted by microarray analysis techniques. The investigations on the expression condition of a large number of genes were carried out by many researchers especially on organism's entire genome. Microarray gene expression data have only few number of samples (less than hundreds), while number of features (attributes) is typically in thousands. Having so many features relative to very few samples creates a high probability of resulting in false positives while trying to spot very important genes. A pediatric/childhood tumor is the most threatening illness that occurs between birth and 14 years. Childhood leukemia (34%), brain tumors (23%), and lymphomas (12%) are the most common cancers that impinge on children. The number of new cases was highest among the 1–4 age groups, but the number of deaths was highest among the 10–14 age groups. It is necessary to diagnose this illness early, in order to improve the survival rate among children. Data analysis is one of the most essential computational tasks when it comes to exploring gene expression data. It includes gene selection, classification and clustering [1]. Gene (feature) selection methods can be used to recognize and get rid of unwanted, irrelevant and redundant features from big microarray gene expression data that do not contribute to the accuracy of classification and prediction models. In other words gene selection methods search for best subset of genes in thousands of gene expression data.

Gene (feature) selection method includes two phases: Feature Selection and Model Construction with performance evaluation [4]. The feature selection phase [5] consists of three steps: (a) generation of candidate set containing a subset of the original features via certain search strategies; (b) evaluating the candidate set and estimating the efficacy of the features in the candidate set. Based on the evaluation, some features in the candidate set may be discarded or added to the selected feature set according to their relevance; and (c) determining whether the current set of selected features are good enough using certain stopping criterion. If it is, a feature selection algorithm will return the set of selected features, otherwise, it iterates until the stopping criterion is met [6, 7]. The most important and prominent features can be selected using either forward selection or backward elimination. Forward selection method starts with an empty set of features, adds one feature at a time and evaluates the importance of features by constructing a classifier model. This process continues until predictive feature subset with higher classification accuracy is achieved. Backward elimination method starts with all features at first and eliminates one feature in each iteration; evaluate features using classifier model and stops when classifier yields higher accuracy. Once the best subset of features was selected, it could be related to revise the training and test data for model fitting and prediction [8]. Table 1 provides a review of feature selection algorithms.

Feature selection may include class labeled training data or unlabeled/partially labeled training data, leading to the development of supervised, unsupervised and semi-supervised feature selection algorithms [9]. In the evaluation process, a supervised feature selection algorithm determines features' significance by evaluating their association with the class or their effectiveness for achieving accurate prediction. Without class labels, an unsupervised feature selection algorithm may make the most of data variance or data distribution in its evaluation of features' significance [10].

Table 1 Review of feature selection methods [9]

Model search	Advantages	Disadvantages	Examples
Univariate filter	Fast, scalable, independent of classifier	Ignores feature dependencies, ignores interaction with classifier	χ^2, information gain, gain ratio, Euclidean distance, i-test
Multivariate filter	Models feature dependencies, independent of classifier, better computational complexity than wrapper methods	Slower and less scalable than univariate techniques, ignores interaction with classifier	Correlation based feature selection (CFS), Markov blanket, fast correlation based feature selection (FCBF)
Deterministic wrapper	Simple, interacts with classifier, models feature dependencies, less computationally expensive than randomized methods	Risk of overfitting, more prone than randomized methods to getting stuck in local optimum, classifier dependent selection	Sequential forward selection, sequential backward elimination, take away r, Plus-q, beam search
Randomized wrapper	Less prone to local optima, interacts with classifier, models feature dependencies	Computationally intensive, classifier dependent selection, higher risk of overfitting than deterministic algorithms	Simulated annealing, randomized hill climbing, genetic algorithms, estimation of distribution algorithms
Embedded methods	Interacts with classifier, models feature dependencies, better computational complexity than wrapper methods	Classifier dependent selection	Decision trees, Weighted Naïve Bayes, feature selection using the weight vector of SVM

A semi-supervised feature selection algorithm uses a small amount of class labeled data as additional information to improve unsupervised feature selection. Depending on how and when the usefulness of selected features is evaluated, different strategies can be adopted, which broadly fall into three categories: filter, wrapper and embedded models [9].

Wrapper model selects features according to the learning algorithm which could result in higher accuracy. The filter model is independent of any learning model, and hence is not biased towards any learning models. Advantage of the filter model is that it allows the algorithms to have very simple structure, which usually employs a straightforward search strategy, such as backward elimination or forward selection, and a feature evaluation method designed according to certain criterion. Embedded method combines positive properties of wrapper and filter method. From the literature, it was perceived that gene selection methods like Chi-square test, Information Gain, QuickReduct and ReliefF yield better performance when applied to medical datasets [11].

Microarray analysis techniques have provided biologists with the facility to measure the expression levels of thousands of genes. This leads to statistical and analytical challenges due to enormous amount of raw gene expression data and including accurate categorization of data. Classification of microarray gene expression data aims at identifying the differentially expressed genes that contribute to categorize the class membership for new samples. Major intricacy in microarray classification is presence of very small number of samples in comparison to the large number of genes in the data [11]. Investigating this high dimensional nature of gene data involves time, resources and labor. In order to cater to the high dimensional gene data, gene selection algorithms can also be executed in parallel to single out differentially expressed genes in very less time. Parallelization of algorithms however leads to certain critical issues like synchronization, communication overhead and load imbalance. In order to address these issues, Hadoop Map Reduce—a parallel programming framework has been used in literature [12, 13]. Hadoop exploits commodity machines to process very large datasets in a cluster of worker nodes. Hadoop consists of a Map Reduce engine that performs parallel computation and a Hadoop Distributed File System (HDFS) for distributed storage. There is a latency in accessing files or results from HDFS which degrades the performance of parallel computation. Spark is an Apache framework which runs on top of YARN (Yet Another Resource Negotiator) and it has the advantage of in-memory computation. Spark provides Machine Learning library (Spark MLlib) which helps to handle execution of machine learning algorithms on huge amount of data [14, 15]. SparkMLlib provides support for various machine learning algorithms to be computed in parallel by utilizing many cores in a processor or group of worker nodes in cluster of machines. In contrast, Weka is an open source data mining software suite [16], which applies only serial computation of data. The *spark.mllib* package is the primary Application Programming Interface (API) for MLlib. The list of machine learning algorithms provided by SparkMLlib is presented in Table 2 [17].

Table 2 List of packages supported by *spark.mllib* for diverse machine learning algorithms

Machine learning algorithm	Supported methods	Packages provided by *spark.mllib* [17]
Binary classification	Linear SVM	classification.{SVMModel, SVMWithSGD} evaluation.BinaryClassificationMetrics util.MLUtils
	Logistic regression	classification.{LogisticRegressionModel, LogisticRegressionWithLBFGS} evaluation.MulticlassMetrics regression.LabeledPoint util.MLUtils
	Decision trees	tree.DecisionTree tree.model.DecisionTreeModel util.MLUtils
	Random forests	tree.RandomForest tree.model.RandomForestModel util.MLUtils
	Gradient-boosted trees	tree.GradientBoostedTrees tree.configuration.BoostingStrategy tree.model.GradientBoostedTreesModel util.MLUtils
	Naïve Bayes	classification.{NaiveBayes, NaiveBayesModel} util.MLUtils
Multiclass classification	Logistic regression, decision trees, random forests, Naïve Bayes	The packages provided are similar to the packages available for binary classification
Regression	Decision trees, random forests, gradient-boosted trees	The packages provided are similar to the packages available for binary classification
	Linear least squares, lasso, ridge regression	linalg.Vectors regression.LabeledPoint regression.LinearRegressionModel regression.LinearRegressionWithSGD regression.StreamingLinearRegressionWithSGD
	Isotonic regression	regression.{IsotonicRegression, IsotonicRegressionModel} util.MLUtils
Collaborative filtering (Recommender system)	Alternating least squares (ALS)	recommendation.ALS recommendation.MatrixFactorizationModel recommendation.Rating
Clustering	K-Means	clustering.{KMeans, KMeansModel} linalg.Vectors
	Gaussian mixture	clustering.{GaussianMixture, GaussianMixtureModel}

(continued)

Table 2 (continued)

Machine learning algorithm	Supported methods	Packages provided by *spark.mllib* [17]
	Power iteration clustering (PIC)	clustering.PowerIterationClustering
	Latent Dirichlet allocation (LDA)	clustering.{DistributedLDAModel, LDA}
	Bisecting k-means	clustering.BisectingKMeans linalg.{Vector, Vectors}
	Streaming k-means	clustering.StreamingKMeans regression.LabeledPoint spark.streaming.{Seconds, StreamingContext}
Dimensionality reduction	Singular value decomposition (SVD)	linalg.Matrix linalg.{Vector, Vectors} linalg.SingularValueDecomposition linalg.distributed.RowMatrix
	Principal component analysis (PCA)	linalg.Matrix linalg.Vectors linalg.distributed.RowMatrix feature.PCA regression.LabeledPoint spark.rdd.RDD
Feature extraction and transformation	Term frequency-inverse document frequency (TF-IDF)	feature.{HashingTF, IDF} linalg.Vector spark.rdd.RDD
	Word2Vec (vector representation of words)	feature.{Word2Vec, Word2VecModel}
	Model fitting	feature.{StandardScaler, StandardScalerModel} linalg.Vectors util.MLUtils
	Normalizer	feature.Normalizer util.MLUtils
	ChiSqSelector	feature.ChiSqSelector linalg.Vectors regression.LabeledPoint util.MLUtils
	ElementwiseProduct	feature.ElementwiseProduct linalg.Vectors
	Principal component analysis (PCA)	feature.PCA linalg.Vectors regression.{LabeledPoint, LinearRegressionWithSGD}

(continued)

Table 2 (continued)

Machine learning algorithm	Supported methods	Packages provided by *spark.mllib* [17]
Frequent pattern mining	FP-growth	fpm.FPGrowth spark.rdd.RDD
	Association rules	fpm.AssociationRules fpm.FPGrowth.FreqItemset
	PrefixSpan	fpm.PrefixSpan
Optimization	Gradient descent and stochastic gradient descent (SGD)	optimization.SGD
	Limited-memory Broyden–Fletcher–Goldfarb–Shanno (L-BFGS)	classification.LogisticRegressionModel evaluation.BinaryClassificationMetrics linalg.Vectors optimization.{LBFGS, LogisticGradient, SquaredL2Updater} util.MLUtils

Preliminary investigations on gene expression data revealed the fact that Chi-square test performed efficiently and accurately in predicting gene data but parallelizing the process of Chi-Square test based gene selection posed impediments in real-time execution. This motivated the authors to propose gene selection approach termed the parallel Chi-Square selector on Apache Spark to identify subsets of prominently expressed genes that are pertinent for predicting cancer. The selected genes were used to build a classification model using parallel Decision Tree and parallel Random Forest with Apache SparkMLlib. The constructed model was evaluated using 10-fold Average Split method and accuracy of classification was analyzed for different sets of selected genes.

Cancer dataset for five different cancer types were gathered from Artificial Intelligence (A.I.) Orange labs, Ljubljana [18]. The microarray gene expression datasets considered in this research work were as follows. 1. Brain Tumor, 2. Gastric Cancer, 3. Glioblastoma, 4. Lung Cancer (2 class and 5 class) and 5. Childhood Leukemia (2 class, 3 class and 4 class). Previous work in this area revealed that gene selection and classification algorithms were executed on a single processor (in a serial fashion) using Weka [16] data mining suite. Authors attempt to parallelize gene (feature) selection and classification algorithms using Apache SparkMLlib in order to scale for high dimensional nature of microarray gene data. The results reveal the fact that classifying gene data without gene selection yields less accuracy when compared to performing gene selection followed by classification. The obtained results were compared with previously reported accuracy for each cancer type from AI Orange labs, Ljubljana. The proposed parallelized gene selection and classification yielded higher accuracy than the previously reported one as discussed in the Results and Discussion section.

2 Feature Selection on Gene Expression Data

Feature Selection approaches on mining gene expression data is concisely presented in the following section. Feature selection uses the two methods to detect pertinent feature for identify the target class label. The two methods are (i) Feature Subset Selection method and (ii) Ranking method.

1. **Feature Subset Selection**

Subset selection methods aspire to pick the best subset of features based on the search space, search method and the filtering criterion [9]. Two most widely used feature subset evaluators are as follows.

a. **Correlation Feature Subset Evaluator (CFS)**

The Correlation Feature Subset (CFS) hypothesis [19] recommend that the most predictive features needed to be highly correlated to the target class and has least correlation with other predictor features.

b. **Fuzzy Rough Subset Evaluator (FRS)**

Fuzzy Rough Subset Evaluator (FRS) method applies Best Fist search approach to identify the minimal subset of features [20]. However, the size of the subset is dependent upon the search space and the search method.

2. **Feature Selection by Ranking**

Ranking feature selection methods [21] are based on certain selection measures and ranks each feature based on its importance in accurate prediction. Five feature selection measures were studied to explore their applicability to select optimal features in the medical domain.

a. **Information Gain**

Shannon [22] first introduced Information Gain (IG) attribute selection measure on information theory that revealed the value or information content of messages. The attribute selected by the information gain minimizes the information needed to classify the tuples in the resulting partitions and reflects the least randomness in these partitions. This approach reduces the expected number of tests needed to classify a given tuple. The expected information needed to classify a tuple in data partition 'D' is given by Eq. 1.

$$\text{Info}_D = - \sum_{i=1}^{m} p_i \log_2 p_i \tag{1}$$

where p_i is the probability that an arbitrary tuple in a data partition 'D' belongs to a class C_i and is estimated by $|C_i, D|/|D|$. A log function to the base 2 is used since the information is encoded in bits. Info_D is just the average amount of information needed to identify the class label of a tuple in D. This is based entirely on the proportions of the tuples in each class. Info(D) is also known as the entropy of D [10].

In order to partition the tuples in D on a discrete-valued attribute, $A\{a_1, a_2, a_v\}$, corresponded directly to all possible outcomes of a test on A. Attribute A could then be used to split 'D' into 'v' partitions or subsets, $\{D_1, D_2, D_v\}$, where D_j contained those tuples in D that had an outcome a_j on A. However, in order to obtain an exact classification, $Info_A(D)$ is calculated which is defined by Eq. 2.

$$\text{Info}_A(D) = \sum_{j=1}^{v} \frac{|D_j|}{D} X \, \text{info}(D_j) \tag{2}$$

$Info_A(D)$ is the expected information required to classify a tuple from a data partition D based on the partitioning by attribute 'A'. The smaller the expected information required, the greater the purity of the partitions. Information gain is defined as the difference between the original information requirement (i.e., based on just the proportion of classes) and the new requirement (i.e., obtained after partitioning on A). That is,

$$\text{Information Gain (A)} = \text{Info}_A - \text{Info}_A(D) \tag{3}$$

The attribute A with the highest information gain, (*Information Gain (A)*), is chosen as the splitting attribute. The drawback of the information gain measure lies in the fact that it is biased toward tests with many outcomes. That is, it prefers to select attributes having a large number of values. In order to overcome this, an extension to this measure was proposed as described in the following section.

b. **Gain Ratio**

Gain Ratio (GA) uses an extension to information gain that attempts to overcome the bias on attributes selected by the information gain criterion. It applies a kind of normalization to information gain using a split information value defined analogously with $Info_A(D)$ as stated by Han and Kamber [10].

$$\text{SplitInfo}_A(D) = -\sum_{j=1}^{v} \frac{|D_j|}{|D|} X \log_2\left(\frac{|D_j|}{|D|}\right) \tag{4}$$

This value represents the potential information generated by splitting the training data set, 'D', into 'v' partitions, corresponding to the 'v' outcomes of a test on attribute 'A'. The number of tuples having a certain outcome with respect to the total number of tuples in 'D' alone was considered. It differs from information gain, which measures the information with respect to classification that is acquired based on the same partitioning [23]. The gain ratio is defined by Eq. 5.

$$\text{GainRatio}_A = \frac{Gain_A}{Split\,Info_A} \tag{5}$$

The attribute with the maximum gain ratio is selected as the splitting attribute. However when the split information approaches 'zero', the ratio becomes unstable. A constraint is added to avoid this, whereby the information gain of the test selected must be large, at least as great as the average gain, over all tests examined.

c. Symmetric Uncertainty

Another suitable approach to select features is based on the Symmetric Uncertainty (SU) values between the feature and the target classes [24]. Symmetry is a desired property for a measure of correlations between features. However, information gain is biased in favour of features with more values. Furthermore, the values have to be normalized to ensure they are comparable and have the same affect. Therefore, symmetrical uncertainty measure was utilized and is defined as follows.

$$SU[X, Y] = 2\left[\frac{IG(X|Y)}{H(X) + H(Y)}\right] \tag{6}$$

It compensates for information gain's bias towards features with more values and normalizes its values to the range [0, 1] with the value '1' indicating that knowledge of the value of either one completely predicts the value of the other and the value '0' indicating that X and Y are independent. In addition, it still treats a pair of features symmetrically. The entropy $H(X)$ of variable X is determined by Eq. 7.

$$H(X) = -\sum_i P(X_i)x \log_2 P(X_i) \tag{7}$$

This entropy-based measures require nominal features, but they can be applied to measure correlations between continuous features as well, if the values are discretized in advance [9].

d. Chi-Square Significance

Chi-squared attribute evaluation evaluates the worth of a feature by computing the value of the chi-squared statistic with respect to the class [25]. The initial hypothesis H_0 is the assumption that the two features are unrelated, and it is tested by the chi-squared formula:

$$\chi^2 = \sum_{i=1}^{c}\sum_{j=1}^{r}\frac{(O_{ij} - e_{ij})^2}{e_{ij}} \tag{8}$$

where O_{ij} is the observed frequency and e_{ij} is the expected (theoretical) frequency, asserted by the null hypothesis. The greater the value of χ^2, the greater will be the evidence against the hypothesis H_0. The chi-square test is the common statistical test that measures divergence from the distribution expected if one assumes the feature occurrence to be independent of the class value [9]. As a statistical test, it is known to behave erratically for very small-expected counts.

e. **ReliefF**

The original Relief algorithm [26] can deal with nominal and numerical attributes. However, the algorithm was unable to process incomplete data and was limited to two-class problems. ReliefF feature selection method is an extension of Relief and it overcomes the drawback of Relief algorithm. ReliefF attribute evaluation, evaluates the worth of a feature by repeatedly sampling an instance and considering the value of the given feature for the nearest instance of the same and different class [9]. This attribute evaluation assigns a weight to each feature based on the ability of the feature to distinguish among the classes, and then selects those features whose weights exceed a user defined threshold as relevant features. The weight computation is based on the probability of the nearest neighbors from two different classes having different values for a feature and the probability of two nearest neighbors of the same class having the same value for the feature. The higher the difference between these two probabilities, the more significant is the feature. Inherently, the measure is defined for a two-class problem, which can be extended to handle multiple classes, by splitting the problem into a series of two-class problems.

From the above discussed feature selection methods using ranking approach, Information Gain is biased towards features with more values. Gain Ratio and Symmetric Uncertainty normalizes the results of Information Gain. But Symmetric Uncertainty works well only for nominal values. ReliefF is more suitable for binary class datasets. Therefore Chi-Square test ranking method was chosen for selecting optimal genes from microarray gene expression data as it is more suitable for choosing relevant genes from continuous values and can be applicable for multi class dataset. The following section briefs about feature selection methods applied on microarray gene data from literature.

Alonso-González et al. [27] proposed the concept of performing attribute selection along with the classification algorithm. The authors' idea was to choose a suitable fusion of feature selection and classifying algorithms for improved classification accuracy while applying these methods on microarray gene expressions data. They have conducted experiments on 30 cancer datasets and selected genes in three different intervals. (i) Genes up to 25 (ii) Gene set ranging from 30 to 140 and (iii) Gene set more than 140. Zhang et al. [28] introduced a new method for gene selection called chi-square test-based integrated rank—gene and direct classifier. Initially weighted integrated rank of gene importance was obtained from chi-square tests of single and pair-wise gene interactions. Then ranked genes were introduced sequentially and redundant genes were removed using leave one out cross validation of chi square test based direct classifier. This was performed within the training set to obtain informative genes. Finally accuracy was measured for independent test data with learned chi-square test based direct classifier. Begum et al. [29] proposed an ensemble based SVM called AdaBoost Support Vector Machine (ADASVM) for classifying cancer from microarray gene expression data. Leukemia dataset was used to compare the performance of ADASVM, K-Nearest Neighbor (K-NN) and Support Vector Machine (SVM) classifiers. Jeyachidra et al. [30] compared feature selection algorithms like Gini Index, Chi square and MRMR (Maximum Relevance and Minimum Redundancy) and evaluated J48 and Naive Bayesian classifier using leave one out cross validation. It was identified that when top 20 genes were selected using Chi square and Gini Index, both J48 and Naive Bayesian classifier yielded classification accuracy of 88%. When MRMR gene selection method was used, Naive Bayesian classifier yielded 83.8% and J48 yielded 85%. Weitschek et al. [31] proposed a novel pipeline-able tool called GELA (Gene Expression Logic Analyzer) to perform a knowledge discovery process in gene expression profile data of RNA-seq. They had tested knowledge extraction algorithm on the public RNA-sequence data sets of Breast Cancer and Abdomen Cancer, along with the public microarray data sets of Psoriasis and Multiple Sclerosis. Cabrera et al. [32] proposed a system that utilizes gene expression data from oligonucleotide microarrays to predict the presence or absence of lung cancer, predict the specific type of lung cancer—should it be present, and determine marker genes that are attributable to the specific kind of the disease. Das et al. [57] had explored classification and clustering techniques for early detection of diabetes. J48 and Naïve Bayesian classifiers were used to predict diabetes at early stage and will assist in timely and proper treatment to patients. Das et al. [58] had proposed a PSO-based evolutionary multilayer perceptron for classification problem and compared with Multilayer Perceptron (MLP) and Genetic Algorithm based Multilayer Perceptron (GA-MLP). The proposed PSO-based MLP performed better than the other two models. A Health Recommendation System (HRS) was proposed by Sahoo et al. [59] that utilize huge volume of clinical data to provide recommendations for kind of treatment to be given to patients. Dey et al. [60] have proposed an Intelligent Healthcare Management system by exploiting big data analytics. Epidemiologic studies have shown that genetic variability is among the factors that affect a person's susceptibility to lung cancer. Support Vector Machine classifier was used to classify lung cancers in LibSVM tool. Lung cancer can be categorized as (a) Adenocarcinoma (ADCA)—further into subclasses C1, C2, C3, C4, CM, GRP1, (b) Squamous Cell Carcinoma (SQ) and (c) Pulmonary Carcinoids (COID). The different gene selection methods applied on gene expression data and its performance in terms of accuracy is presented in Table 3. The accuracy of classification algorithm was obtained by evaluating the classifier using k-fold cross validation.

3 Parallel Programming Framework: Apache Hadoop and Spark

The machine learning methods used in bioinformatics are iterative and parallel. The distributed and parallel computing technologies would help in handling big data. Usually big data tools perform computation in batch-mode and are not optimized for iterative processing and high data dependency among operations. In the recent years, parallel, incremental, and multi-view machine learning algorithms have been proposed. Similarly, graph-based architectures and in-memory big data tools have been developed to minimize I/O cost and optimize iterative processing [38]. Some

Table 3 Review of gene selection algorithms on microarray gene datasets

Authors (year)	Title	Microarray gene dataset	Gene selection algorithm	Classification algorithm	Accuracy in percentage
Zhang et al. [28]	Informative gene selection and direct classification of tumor based on chi-square test of pair wise gene interactions	Leukemia1 Leukemia2 Breast cancer Lung cancer	Chi-square test-based integrated rank gene and direct classifier (χ^2-IRG-DC)	Naive Bayes, K nearest neighbor, SVM and χ^2-DC	85.86
Nguyen et al. [33]	Random forest classifier combined with feature selection for breast cancer diagnosis and prognostic	Wisconsin Breast cancer Dataset	New feature ranking formula	Random forest	99.8
Rajeswari et al. [34]	Feature selection for classification in medical data mining	Breast cancer	Genetic search feature selection with decision tree classifier	Neural network	77.23
				Decision tree	82.17
				SVM	85.81
Lavanya et al. [35]	Ensemble decision tree classifier for breast cancer data	Breast cancer Wisconsin (diagnostic)	Symmetric UncertAttributesetEval	Random forest	95.96
Umpai and Aitken [11]	Feature selection and classification for microarray data analysis: evolutionary methods for identifying predictive genes	Leukemia (AML and Pre-T ALL leukemia)	Rank gene and evolutionary algorithm	ID3	98.24
Hassanien [36]	Classification and feature selection of breast cancer data based on decision tree algorithm	Breast cancer	Feature subset selection	Decision tree	99.55
Ben-Dor et al. [37]	Tissue classification with gene expression profiles	Colon cancer Ovarian cancer Leukemia	Threshold number of misclassification (TNoM) score	K nearest neighbor, SVM, AdaBoost	90

of the big data tools used for mining data from Gene Expression data was depicted in Table 4.

Parallel Decision Tree on Hadoop Map Reduce will reduce communication cost and execution time. It is also scalable to large data [43]. Decision tree classifier has less error rate and is easy to implement. However the accuracy of decision tree classifier decreases with increase in number of missing values in attributes [44]. Accuracy of parallel decision tree can be improved using an ensemble technique called Random Forest [45–49]. Random Forest is the collection of decision trees used to predict the target class label of the test data. Richards et al. had reported that Random Forest performs well compared to other classification algorithms viz Neural Network, SVM and Decision Tree, especially on high-dimensional data [50]. Table 5 presents a brief review on parallelized gene selection and classification algorithms applied on big datasets. This triggered a thought of utilizing parallel gene selection and classification algorithms on huge microarray gene data.

Based on the review of previous work in the area of gene selection on microarray gene expression data, it was identified that (i) There is a clear scarcity of efficient parallel algorithms to mine from voluminous gene data. (ii) Parallelization of gene selection approaches has not been explored on clinical data although its importance has been emphasized in the literature. (iii) There is a need to improve classification accuracy when genes are selected in parallel on Hadoop Map Reduce. (iv) Cancer is one of the major causes for mortality, especially in low and middle-income countries. Yet, not much work has been devoted to unearthing the genetic variants contributing to the disease.

This being the rationale behind this research, the authors have made an investigation on the possibility of mining optimal genes contributing to five different cancers through computational methods based on parallelization of gene selection and classification techniques. The dataset description, methodology and experimental results are discussed in the following sections.

4 Materials and Dataset Description

Microarray gene expression dataset for five different cancer types has been collected from AI Orange labs, Ljubljana [18]. Datasets include: 1. Brain Tumor, 2. Gastric Cancer, 3. Glioblastoma, 4. Lung Cancer (2 class and 5 class) and 5. Childhood Leukemia (2 class, 3 class and 4 class). The platform from which dataset was collected is Affymetrix Human Genome Array. All gene expression datasets consisted of continuous values. The description of cancer gene expression data is tabulated in Table 6.

Table 4 Big data tools used for analysis of Gene Expression data

Authors (year)	Title	Tool name	Category	Description
Stokes et al. [39]	chip artifact CORRECTion (caCORRECT): a bioinformatics system for quality assurance of genomics and proteomics array data.	(caCORRECT)	Quality assurance tool for bioinformatics	1. It removes artifactual noise from high throughput microarray data 2. caCORRECT may be used to improve integrity and quality of both public microarray archives as well as reproduced data and to provide a universal quality score for validation
Phan et al. [40]	omniBiomarker: a web-based application for knowledge-driven biomarker identification.	omniBiomarker	Biomarker identification tool for Gene Expression data	It uses knowledge-driven algorithms to find differentially expressed genes for biomarker identification from high throughput gene expression data
Li et al. [41]	Sparkbench: a comprehensive benchmarking suite for in-memory data analytic platform spark	Apache Spark	Big data framework	Big data analysis framework which provides fault tolerant, scalable and easy to-use in memory abstraction
Koliopoulos et al. [42]	A parallel distributed Weka framework for big data mining using spark	DistributedWekaSpark	Distributed framework for Weka with user interface	1. The framework is implemented on top of Spark, a Hadoop-related distributed framework with fast in-memory processing capabilities and support for iterative computations 2. By combining Weka's usability and Spark's processing power, *DistributedWekaSpark* provides a usable prototype distributed Big Data Mining workbench that achieves near-linear scaling in executing various real-world scale workloads—91.4% weak scaling efficiency on average and up to $4\times$ faster on average than Hadoop

Table 5 Review of parallelized gene selection and classification algorithms on Hadoop Map Reduce

Authors (year)	Title	Big dataset	Parallelized gene selection method	Parallelized classification algorithm	Accuracy in percentage
Islam et al. [51]	MapReduce based parallel gene selection method	Colon cancer	Between-groups to Within-groups sum of square	Map reduce K nearest neighbour	100
		Leukemia			98
		Lymphoma			85
		Prostate cancer			93
Peralta et al. [52]	Evolutionary feature selection for big data classification: a MapReduce approach	Epsilon	Evolutionary feature selection (gene algorithm)	SVM, logistic regression and Naive Bayes	65–71
		ECBDL14-ROS			50–65

5   Parallel Gene Selection Methodology

5.1   Chi-Square Test Based Gene Selection

Research on previous literature and preliminary investigations on gene data involved exploring the performance of various feature selection approaches viz, Chi-Square test, Information Gain, ReliefF, Fisher Filtering, Qucik Reduct and Symmetric Uncertainity on gene data. This enabled the authors to identify the fact that Chi-Square test yielded the optimal set of predictive genes that enhanced the classification accuracy.

Chi-Square Correlation Coefficient was utilized for finding correlation between genes (attributes) [10]. Chi Square value is computed as follows.

$$\chi^2 = \sum_{i=1}^{c}\sum_{j=1}^{r} \frac{(O_{ij} - e_{ij})^2}{e_{ij}} \quad \text{(Ref. Eq.8)}$$

where O_{ij} is observed (actual) frequency of joint event of genes (A_i, B_j) and 'e_{ij}' is expected frequency of (A_i, B_j) which is computed as follows. The values 'r' and 'c' are number of rows and columns in contingency table.

$$e_{ij} = \frac{Count(A = a_i) \times Count(B = b_j)}{N} \quad (9)$$

Table 6 Microarray Gene Expression Dataset Description

Gene dataset	No. of genes	No. of samples	Target class	Cancer sub-types
Brain tumor	7129	40	5	1. **Medulloblastoma** (medulloblastoma): 10 examples (25.0%) 2. **Malignant glioma** (glioma): 10 examples (25.0%) 3. **Rhabdoid tumor** (RhabdoidTu): 10 examples (25.0%) 4. **Normal cerebellum** (Normal): 4 examples (10.0%) 5. **Primitive neuroectodermal** tumor (PNET): 6 examples (15.0%)
Gastric cancer	4522	30	3	1. **Normal gastric tissue** (Normal): 8 examples (26.7%) 2. **Diffuse gastric tumor** (Diffuse): 5 examples (16.7%) 3. **Intestinal gastric tumor** (Intestinal): 17 examples (56.7%)
Glioblastoma	12625	50	4	1. **Classic glioblastoma (CG)**: 14 examples (28.0%) 2. **Classic oligodendroglioma (CO)**: 7 examples (14.0%) 3. **Nonclassic glioblastoma (NG)**: 14 examples (28.0%) 4. **Nonclassic oligodendroglioma (NO)**: 15 examples (30.0%)
Lung cancer	10541	34	3	1. **Squamous cell carcinoma** (Squamous): 17 examples (50.0%) 2. **Adenocarcinoma** (Adenocarcinoma): 8 examples (23.5%) 3. **Normal lung tissue** (Normal): 9 examples (26.5%)
Lung cancer	12600	203	5	1. **Adenocarcinoma (AD)**: 139 examples (68.5%) 2. **Normal lung (NL)**: 17 examples (8.4%) 3. **Small cell lung cancer (SMCL)**: 6 examples (3.0%) 4. **Squamous cell carcinoma** (SQ): 21 examples (10.3%) 5. **Pulmonary carcinoid** (COID): 20 examples (9.9%)

(continued)

Table 6 (continued)

Gene dataset	No. of genes	No. of samples	Target class	Cancer sub-types
Childhood tumor	9945	23	2	1. **Ewing's sarcoma (EWS)**: 11 examples (47.8%) 2. **Rhabdomyosarcoma (RMS)**: 12 examples (52.2%)
Childhood tumor	9945	23	3	1. **Ewing's sarcoma (EWS)**: 11 examples (47.8%) 2. **embryonal rhabdomyosarcoma (eRMS)**: 3 examples (13.0%) 3. **alveolar rhabdomyosarcoma (aRMS)**: 9 examples (39.1%)
Childhood leukemia (acute lymphoblastic leukemia)	8280	60	4	1. **Mercaptopurine alone (MP)**: 13 examples (21.7%) 2. **High-dose methotrexate (HDMTX)**: 21 examples (35.0%) 3. **Mercaptopurine and low-dose methotrexate (LDMTX_MP)**: 16 examples (26.7%) 4. **Mercaptopurine and high-dose methotrexate (HDMTX_MP)**: 10 examples (16.7%)

where N is number of data tuples. Count $(A = a_i)$ is number of tuples having value a_i for A. Count $(B = b_j)$ is number of tuples having value b_j for B, where 'A' and 'B' represent the gene's (attributes) under evaluation. The sum is computed over all of $r \times c$ cells in a contingency table. The χ^2 value needs to be computed for all pair of genes. The χ^2 statistics test the hypothesis that genes A and B are independent. The test is based on significance level with $(r - 1) \times (c - 1)$ degrees of freedom. If Chi-Square value is greater than the statistical value for given degree of freedom, then the hypothesis can be rejected. If the hypothesis can be rejected, then we say that genes A and B are statistically related or associated [10].

Table 7 presents different gene selection algorithms applied on colon cancer dataset and their highest performance (accuracy) in bold for corresponding classifier. It is evident that Chi-Square test performs better with Decision Tree classifier and obtains highest accuracy as 88.71%.

This was considered as the base idea for further investigations on other gene expression datasets involved in this research work. The following section briefs about the process involved in parallelizing algorithms like Chi-Square test based gene selection approach, Decision Tree and Random Forest Classifiers.

Table 7 Accuracy obtained by gene selection and classification algorithms on colon cancer dataset

Authors (year)	Title	Gene selection algorithm	Gene dataset	Decision tree (%)	Naïve Bayes (%)	SVM (%)	K-NN (%)
Wang and Gotoh [53]	A robust gene selection method for microarray-based cancer classification	α-depended degree-based feature selection	**Colon cancer**	**88.71**	91.93	**88.71**	**88.71**
		Chi-square test		**88.71**	90.32	87.10	87.10
		Information gain		85.48	85.48	87.10	87.10
		Relief-F		87.10	85.48	87.10	87.10
		Symmetric uncertainty		87.10	**91.94**	87.10	**88.71**

5.2 Parallelized Chi-Square Selector for Gene Selection

Gene expression data need to be partitioned into 'n/p' partitions, where 'n' relates to the number of genes and 'p' represents the number of processors (number of cores in a processor) in a parallel programming environment. The Chi-Square calculation is computed by Map and Reduce functions in each processor.

Mapper Task: Computes the ratio of the squared difference of actual (observed) frequency deviated from expected frequency.

Reducer Task: Computes the summation for calculating χ^2 value.

Finally if χ^2 value is greater than the statistical value for a given degree of freedom, then the genes are considered to be strongly correlated for disease classification. Those genes were ranked highest by Chi Square selector and considered for further investigations.

Algorithm for Parallelized Chi-Square Selector

Input: Microarray Gene Expression data with thousands of genes.

Output: Optimal Genes for cancer classification and prediction.

Algorithm:

1. For each gene pair in gene expression data do
   a. **Map Task**
      i.   Compute Expected Frequency
      $$e_{ij} = = \frac{Count\left(A = a_i\right) \times Count\left(B = b_j\right)}{N}$$
      ii. Compute $T = \dfrac{\left(o_{ij} - e_{ij}\right)2}{e_{ij}}$

   b. **Reduce Task**
      i.   For each r and c, compute $\chi 2$ value
      $$\chi^2 = \Sigma_{i=1}^{c} \Sigma_{j=1}^{r} \frac{\left(o_{ij} - e_{ij}\right)2}{e_{ij}}$$

2. If χ^2 value is greater than statistical value, then those genes are selected for classification.
3. Return selected genes.

5.3 *Parallelized Gene Selection and Classification (PGSC)*

The authors implemented parallelized gene selection and classification for predicting cancer subtypes based on the optimal set of predictive genes. Although Chi-Square test has been utilized in the past for gene selection, parallelizing the approach on gene data posed hurdles in implementation. Apache Spark Machine Learning Library (Spark MLlib) framework was used for execution of parallel Chi-square gene selector and parallel Decision Tree and Random Forest classifiers [54]. Spark provided in-memory computation compared to Hadoop Map Reduce. Hence, the overall execution time on Spark was much less compared to Hadoop Map Reduce [55]. The following procedure outline the steps involved in applying parallelized gene selection and classification algorithms on Gene Expression data for categorizing subtypes of five cancer types.

1. Microarray gene expression data for five different cancer types were collected from AI Orange labs, Ljubljana [18].
2. Parallel Chi-Square Selector (ChiSqSelector) was executed on Apache Spark in order to determine the most important genes for cancer classification. Chi-Square test based ranking method was identified as a more suitable gene selection algorithm for continuous values and is not biased towards values of the genes.
3. Top ranked genes were selected in different variations starting from 1000, 900, 700, 500, 200, 100, 50 to 25, for each class of dataset. The selection of the top ranked gene subset size was done based on the method presented in [27].
4. Selected genes were to be evaluated by classification algorithms like Parallel Decision Tree and Parallel Random Forest on Spark. It was identified that Decision Tree does not consider all the genes for constructing the training model. It selects only important genes. Random Forest forms an ensemble of Decision trees to classify cancer subtypes. In this work authors have considered 150 Decision Trees to form a Random Forest.
5. Finally the prediction accuracy was evaluated using the following metrics: Precision, Recall and F1 Score. The authors have evaluated classifier accuracy using 10-fold Average Split method which takes an average of 10 accuracy values obtained using 80% training and 20% test dataset. The algorithm for executing Decision Tree and Random Forest on Hadoop Map Reduce is given below.

Algorithm for Decision Tree on Hadoop Map Reduce

Algorithm: Random Forest (D, d) - Train a decision tree model from random samples of training instances.

Input: Dataset D, subspace dimension d

Output: Decision Tree Model

Training Phase

1. Split the dataset as 80% for training and 20% for testing.

2. The training dataset is partitioned into multiple subsets of 64 MB or 128 MB and each dataset are given to a Mapper.

3. Map Task

For each subset of data in training dataset, each core in a processor does the following.

    i. For each data attribute, collect class distribution information of the local data.

    ii. Mapper outputs *<attribute, frequency of attribute>* as key value pair for its local data.

4. Reduce Task

    i. Reducer will sum the frequency of attributes and finds best attribute using Entropy. This attribute forms the root of the decision tree.

    ii. Reducer splits training samples based on splitting criteria and writes to Hadoop Distributed File System (HDFS).

5. Load the dataset from HDFS on the Mapper again (overcome by Spark) and repeat steps from 2 to 4 until there are no samples to assign to a class / all samples belong to a class.

Prediction Phase

6. Test data to be given to the constructed decision tree model.

7. Decision tree predicts class label for given test data and emits <test_data , class label> as key value pair.

8. Return predicted class label for test data.

Algorithm for Random Forest on Hadoop Map Reduce

Algorithm: Random Forest (D, T, d) - Train an ensemble of decision tree models from random samples of training instances.

Input: Dataset D, No. of trees T, subspace dimension d

Output: Ensemble of decision trees whose prediction is to be combined by voting or averaging.

1. for t = 1 to T do

    i. Build a bootstrap sample D_t from D by sampling |D| data points with replacement.

    ii. Select 'd' features at random and reduce dimensionality of D_t accordingly.

    iii. Map Task

        Construct a decision tree model M_t from D_t

2. Return $\{Mt \mid 1 \le t \le T\}$

3. Test data to be given to each of the constructed decision tree model.

4. Each decision tree predicts class label for given test data and emits <tree_id , class label> as key value pair.

5. Reduce Task

    i. Reducer will count each class label and computes majority of class.

    ii. Reducer assigns test data to majority class.

6. Return predicted class label for test data.

5.4 Evaluation Method: 10-fold Average Split Accuracy

The entire gene dataset has been split as 80% for training and 20% for testing. As this percentage split uses random sampling technique on finalizing training and testing data, the accuracy of classification keeps varying. So there is a need for evaluating the constructed classification model. The authors explored an evaluation method called 10-fold Average Split. In this method, each gene dataset was split as 80% for training and 20% for testing. Parallel Classification models have been constructed using training dataset and evaluated with testing dataset. The above method has to be repeated for N iterations (say 10 here) and accuracy obtained during every iteration was taken into account. Finally, the average of obtained accuracy was computed as detailed below in the algorithm. Let 'N' be the number of iterations and let '$Accu_i$' be the accuracy obtained in each iteration.

Algorithm for N-fold Average Split

Input: Microarray Gene Expression data with thousands of genes.

    N- Number of iterations

Output: Average Accuracy on percentage split.

Algorithm:

1. For N number of iterations
2. Split each gene dataset as 80% for training and 20% for testing.
       iii.   Construct a model on training dataset using parallelized classification algorithm.
       iv.   Evaluate the constructed model using test dataset.
       v.   Obtain accuracy of parallel classification as $Accu_i$
3. Repeat step 2 for N number of iterations.
4. Compute the Average Split accuracy as follows.

$$\text{Avg-Split Accuracy} = \frac{1}{N}\sum_{i=1}^{N} Accu(i)$$

5. Return Avg-Split Accuracy.

---

6 Results and Discussion

Datasets for five tumor types were collected from A.I. Orange labs, Ljubljana [18] and dataset details are briefed in Table 6. Apache Spark environment was established on Ubuntu Operating System and the data was processed on Spark using SparkMLlib. Once the data was collected, Parallel Gene Selection algorithm, Chi-Square Selector was executed in order to select genes that have a high predictive power in predicting the presence of tumours. Parallel Chi-Square Selector was executed to select top ranked genes in gene set sizes ranging from 1000, 900, 700, 500, 200, 100, 50 and 25. Parallel Decision Tree and Parallel Random Forest was implemented on Spark and a classifier model was constructed for each of these gene groups. Accuracy of classifier for each gene group was measured using metrics like Precision, Recall and F1 score. From the entire dataset, 80% of samples were randomly chosen as training and 20% of samples as test instances. Furthermore, the constructed model was evaluated using 10-fold Average Split method, which computes average accuracy of percentage split for 10 iterations. The same was implemented for serial (non-parallelized) execution and both the results were compared with previously reported accuracy given by A.I. Orange labs.

Figures 1, 2, 3, 4, 5, 6, 7, and 8 portrays the accuracy of Chi-Square test based classification for serial and parallel execution of eight microarray gene expression datasets. Table 8 presents the list of abbreviations used in this work.

Figure 1 shows the accuracy of Chi-Square test based serial and parallel execution of Decision Tree and Random Forest, for prediction of Brain tumor with five diagnostic classes. Initially the accuracy of serial and parallel execution of Decision Tree and Random Forest was less when full gene set was used for categorizing cancer subtypes. Further Chi-Square test based gene selection algorithm was used to select

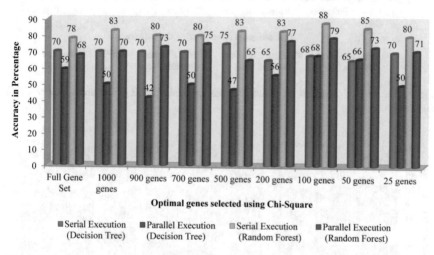

Fig. 1 Accuracy of classifiers on varied gene sets for prediction of Brain tumors with 5 diagnostic classes

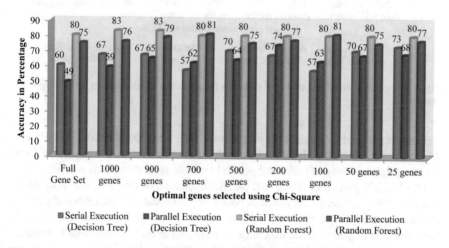

Fig. 2 Accuracy of classifiers on varied gene sets for prediction of gastric tumors with 3 diagnostic classes

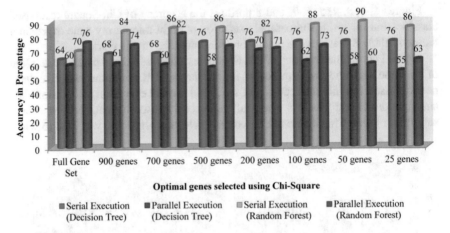

Fig. 3 Accuracy of classifiers on varied gene sets for prediction of glioblastoma tumors with 4 diagnostic classes

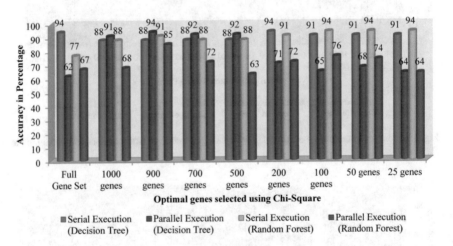

Fig. 4 Accuracy of classifiers on varied gene sets for prediction of lung tumors with 3 diagnostic classes

top ranked gene sets with sizes ranging across 1000, 900, 700,500, 200, 100, 50, and 25, following which the classification model was constructed. It was evident from the chart that accuracy of serial and parallel classification algorithms was higher for top ranked 100 gene sets. Serial and parallel execution of Decision tree resulted in equal accuracy whereas with Random Forest, serial execution gave 88% and parallel execution gave 79% accuracy for optimal set of top 100 genes. On considering all the other sets of genes, serial execution gave higher accuracy compared to parallel execution and this motivates further research in this direction.

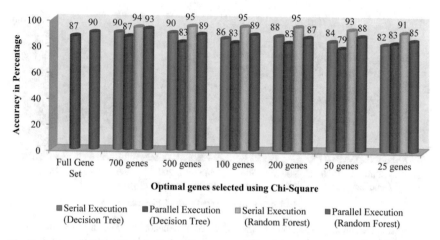

Fig. 5 Accuracy of classifiers on varied gene sets for prediction of lung tumors with 5 diagnostic classes

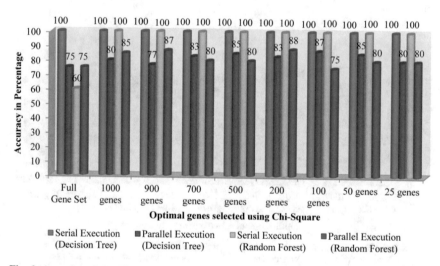

Fig. 6 Accuracy of classifiers on varied gene sets for prediction of childhood tumors with 2 diagnostic classes

Figure 2 shows the accuracy of Chi-Square test based serial and parallel execution of Decision Tree and Random Forest, for prediction of Gastric tumor with three diagnostic classes. Initially the accuracy of serial and parallel execution of Decision Tree and Random Forest was less when full gene set was used for categorizing cancer subtypes. The process of evaluating the varying gene sets that were filtered based on chi-square test was similar for all datasets. It was evident that the accuracy of serial and parallel classification algorithms was higher for top ranked 200 genes. Serial execution of Decision Tree resulted in 67% and parallel execution of Decision

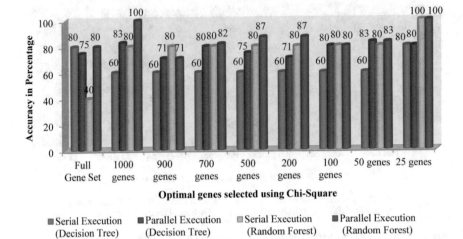

Fig. 7 Accuracy of classifiers on varied gene sets for prediction of childhood tumors with 3 diagnostic classes

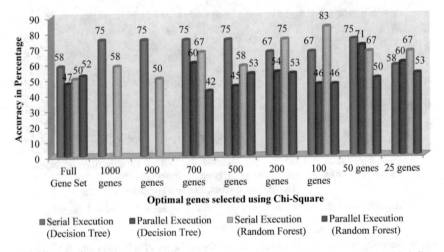

Fig. 8 Accuracy of classifiers on varied gene sets for prediction of childhood tumors with 4 diagnostic classes

tree resulted in 74% accuracy whereas with Random Forest, serial execution gave 80% and parallel execution gave 77% accuracy for optimal set of top 200 genes. On considering all the other set of genes, except gene sets 700, 200, and 100, serial execution gave higher accuracy than parallel execution which needs to be improved.

Figure 3 shows the accuracy of Chi-Square test based serial and parallel execution of Decision Tree and Random Forest, for prediction of Glioblastoma tumor with four diagnostic classes. Initially the accuracy of serial and parallel execution of Decision Tree and Random Forest was less when full gene set was used for categorizing

Table 8 Abbreviation of words

Abbreviation	Description
DT	Decision tree
RF	Random forest
RWFS	Rank weight feature selection
Brain5c	Brain tumor with 5 class
Gastric3c	Gastric cancer with 3 class
Glio4c	Glioblastoma with 4 class
Lung3c	Lung cancer with 3 class
Lung5c	Lung cancer with 5 class
Child2c	Childhood tumor with 2 class
Child3c	Childhood tumor with 3 class
Child4c	Childhood leukemia with 4 class

cancer subtypes. As the dataset was huge (50 samples with 12625 genes), top ranked 1000 genes couldn't be selected using Chi-Square test due to the limited size of the array. It was evident that the accuracy of serial and parallel classification algorithms was higher for top ranked 200 and 100 gene sets. Serial execution of Decision Tree resulted in 76% and parallel execution of Decision tree resulted in 70% accuracy whereas with Random Forest, serial execution gave 82% and parallel execution gave 71% accuracy for optimal set of top 200 genes. On considering all the other set of genes, serial execution gave higher accuracy than parallel execution.

Figure 4 shows the accuracy of Chi-Square test based serial and parallel execution of Decision Tree and Random Forest, for prediction of Lung tumor with three diagnostic classes. Initially the accuracy serial and parallel execution of Decision Tree and Random Forest was less when full gene set was used for categorizing cancer subtypes. It was evident from the chart that accuracy of serial and parallel classification algorithms was higher for top ranked 900 gene set. Serial execution of Decision Tree resulted in 88% and parallel execution of Decision tree resulted in 94% accuracy whereas with Random Forest, serial execution gave 91% and parallel execution gave 85% accuracy for optimal set of top 900 genes. On considering all the other set of genes, serial execution gave higher accuracy than parallel execution which needs to be improved.

Figure 5 shows the accuracy of Chi-Square test based serial and parallel execution of Decision Tree and Random Forest, for prediction of Lung tumor with five diagnostic classes. Initially the accuracy serial and parallel execution of Decision Tree and Random Forest was less when full gene set was used for categorizing cancer subtypes. Serial execution failed with full gene set due to insufficient heap space in the memory as the dataset was massive with 12600 genes and 203 samples. The same dataset could be executed using parallelized algorithms. It was evident from the chart that accuracy of serial and parallel classification algorithms was higher for top ranked 700 gene set. Serial execution of Decision Tree resulted in 90% and parallel execution of Decision tree resulted in 87% accuracy whereas with Random Forest, serial execution gave 94% and parallel execution gave 93% accuracy for optimal set of top 700 genes. On considering all the other set of genes, serial execution gave higher accuracy than parallel execution which needs to be improved.

Figure 6 shows the accuracy of Chi-Square test based serial and parallel execution of Decision Tree and Random Forest, for prediction of Childhood tumor with two diagnostic classes. Initially the accuracy serial and parallel execution of Decision Tree and Random Forest was less when full gene set was used for categorizing cancer subtypes. It was evident from the chart that accuracy of serial and parallel classification algorithms was higher for top ranked 200 gene set. Serial execution of Decision Tree resulted in 100% and parallel execution of Decision tree resulted in 83% accuracy whereas with Random Forest, serial execution gave 100% and parallel execution gave 88% accuracy for optimal set of top 200 genes. On considering all the other set of genes, serial execution gave higher accuracy than parallel execution which needs to be improved.

Figure 7 shows the accuracy of Chi-Square test based serial and parallel execution of Decision Tree and Random Forest, for prediction of Childhood tumor with three diagnostic classes. Initially the accuracy serial and parallel execution of Decision Tree and Random Forest was less when full gene set was used for categorizing cancer subtypes. It was evident from the chart that accuracy of serial and parallel classification algorithms was higher for top ranked 25 gene set. Serial and parallel execution of Decision tree and Random Forest resulted in equal accuracy as 80% and 100% respectively for optimal set of top 25 genes. On considering all the other set of genes, serial execution gave lower accuracy than parallel execution which was a notable and distinct one from all the other tumor datasets considered in this work.

Figure 8 shows the accuracy of Chi-Square test based serial and parallel execution of Decision Tree and Random Forest, for prediction of Childhood tumor with four diagnostic classes. Initially the accuracy serial and parallel execution of Decision Tree and Random Forest was less when full gene set was used for categorizing cancer subtypes. As the dataset was huge (60 samples with 8280 genes), top ranked 1000 and 900 gene set couldn't be selected using Chi-Square test due to limited size of array. It was evident from the chart that accuracy of serial and parallel classification algorithms was higher for top ranked 50 gene set. Serial execution of Decision Tree resulted in 75% and parallel execution of Decision tree resulted in 71% accuracy whereas with Random Forest, serial execution gave 67% and parallel execution gave 50% accuracy for optimal set of top 50 genes. On considering all the other set of genes, serial execution gave higher accuracy than parallel execution which needs to be improved.

Evaluation results were compared with existing work as follows:

1. Accuracy with full set of genes versus accuracy with optimal set of genes obtained by Chi-Square test method.
2. Execution time for classifying cancer subtypes using full gene set and optimal gene set.
3. Previously reported accuracy from A.I. Orange labs versus accuracy obtained from Chi-Square test based serial and parallel Classification.

Table 9 Accuracy with full set of genes versus accuracy with optimal set of genes obtained from chi-square test method

Dataset	Full gene set	Accuracy with full geneset on serial execution		Accuracy with full gene set on parallel execution		Accuracy with optimal gene set on serial execution		Accuracy with optimal gene set on parallel execution		Optimal gene set using chi-square test
		DT	RF	DT	RF	DT	RF	DT	RF	
Brain5c	7129	[a]70	78	59	68	68	88	68	79	100
Gastric	4522	60	80	49	75	67	80	74	77	200
Glio4c	12625	64	70	60	76	76	82	70	71	200
Lung3c	10541	94	77	62	67	88	91	94	85	900
Lung5c	12600	Serial execution failed		87	90	90	94	87	93	700
Child2c	9945	100	60	75	75	100	100	83	88	200
Child3c	9945	80	40	75	80	80	100	80	100	25
Child4c	8280	58	50	47	52	75	67	71	50	50

[a]values to be compared are indicated by the distinguishing colors (red -red/green-green etc.)

4. Previously reported accuracy from Rank Weighted Feature Selection [56] versus accuracy obtained from Chi-Square test based serial and parallel Classification.
5. All the above mentioned results were compared pertaining to both Chi-Square test based serial and parallel Classification.

Table 9 depicts classification accuracy obtained with and without gene selection algorithm. Chi-Square test was executed for all eight datasets and their corresponding optimal number of genes was tabulated in the last column of Table 9. Similarly parallelized Chi-Square test based gene selection algorithm was run on all eight datasets. Model was constructed with parallelized decision tree and random forest using SparkMLlib and it was evaluated with 10-fold Average Split. Moreover when Lung5c dataset (with 12600 genes and 203 samples) was considered, serial execution failed on entire dataset, whereas it is possible to execute the same dataset in parallel on Spark, which utilizes many cores in a processor. On the other hand, when optimal number of genes were selected using Chi-Square test, it became feasible to serially execute lung5c dataset. On comparing serial (Serial) execution with Spark's parallel execution, classification accuracy is more or less similar which motivates the authors to further investigate on improving accuracy for big datasets using parallelized data mining algorithms.

Table 10 shows the variation in execution time (in seconds) for parallelized classification algorithm on Spark. It conveys that execution time for classifying cancer subtypes with optimal gene set is 2-fold lesser than execution time for classifying cancer subtypes with full gene set.

Table 11 depicts the comparison of previously reported accuracy for all eight datasets from AI Orange Labs with Chi-Square test based serial and parallel classification (second column with rest of the columns). The newly reported accuracy was higher for all datasets except lung5c. Lung5c dataset yields equal accuracy as that of previously reported accuracy (highlighted in bold face).

Table 12 depicts the accuracy obtained using Rank Weight Feature Selection based classification and Chi-Square test based serial and parallel Classification.

Table 10 Execution time in seconds for classifying cancer subtypes with and without gene selection

Dataset	Full gene set	Execution time in seconds for full gene set (parallel execution)		Execution time in seconds for optimal gene set (parallel execution)		Optimal gene set using chi-square test
		DT	RF	DT	RF	
Brain5c	7129	15	21	4	7	100
Gastric	4522	8	11	4	6	200
Glio4c	12625	21	33	5	7	200
Lung3c	10541	13	20	5	8	900
Lung5c	12600	36	47	6	11	700
Child2c	9945	9	12	4	5	200
Child3c	9945	12	16	2	3	25
Child4c	8280	29	45	4	7	50

Table 11 Previously reported accuracy from A.I. Orange labs [18] versus accuracy obtained from chi-square test based serial and parallel classification

Dataset	AI Lab's accuracy	Chi-square test based serial classification		Chi-square test based parallel classification		Optimal gene set using chi-square test
		DT	RF	DT	RF	
Brain5c	52.5	68	88	68	79	100
Gastric3c	60	67	80	74	77	200
Glio4c	70	76	82	70	71	200
Lung3c	65.8	88	91	94	85	900
Lung5c	**94**	90	94	87	93	700
Child2c	95	100	100	83	88	200
Child3c	78.3	80	100	80	100	25
Child4c	36.6	75	67	71	50	50

Table 12 Previously reported accuracy from rank weighted feature selection (RWFS) [56] versus accuracy obtained from chi-square test based serial and parallel classification

Dataset	No. of genes using RWFS	RWFS-based classification	Chi-Square test based serial classification	Chi-Square test based parallel classification	Optimal gene set using chi-square test
Brain5c	3	77.5	**88**	**79**	100
Gastric3c	4	93.3	80	81	700
Glio4c	5	90	86	82	700
Lung3c	3	94.1	91	**94**	900
Child4c	6	65	**75**	**71**	50

Rank Weight Feature Selection (RWFS) algorithm [55] presents the minimal and optimal features (genes) by ranking each feature obtained from six feature selection algorithms viz Fuzzy Rough Subset, Information Gain, Symmetric Uncertainty, Chi-Square Co-efficient, ReliefF factor and Information Gain. RWFS algorithm selects genes that are commonly reported by six feature selection algorithms and ranks them based on how many times they are reported by the feature selection algorithms. In

Table 12, columns 2 and 3 represent minimal number of genes reported by RWFS algorithms and their respective classification accuracy. Last column presents the optimal genes obtained using Chi-Square test method. Columns 4 and 5 represent the serial and parallel classification accuracy obtained using Chi-Square test respectively. On comparing classification accuracy of RWFS with classification accuracy of Chi-Square test, Chi-Square test based classification yields higher accuracy for brain5c, lung3c and child4c datasets. This paves way for further investigations on gasric3c and glio4c datasets. The complexity of parallelized feature selection and classification algorithm depends on the number of machines in the Spark environment and amount of job to done by individual machine in a cluster of distributed machines. It also depends on the data complexity.

Moreover, further research is ongoing in order to identify the reason behind the reduced performance when gene selection and classification algorithms are parallelized. Authors believe that if the underlying cause for the inhibition in accuracy is unearthed, it will pave the way to develop improved algorithms that could be generalized to all datasets. Parallel Deep Learning algorithms for feature selection and classification would be explored to predict disease with negligible misclassification rate.

7 Conclusion

Microarray gene expression data has thousands of genes and all genes may not be necessary for classifying and predicting tumors. Gene selection methods aid in enhancing the accuracy of classifiers and in minimizing the training time. A review of gene selection algorithms and parallelized algorithms on Hadoop Map Reduce to scale for big GE datasets were presented. As microarray gene data is growing in size, parallelized algorithms were explored using SparkMLlib. Parallel Chi-Square Selector was executed on Apache SparkMLlib. Genes were selected in varying subset sizes and selected genes were utilized to construct a model using parallel Decision Tree and Random Forest classifiers. The results were evaluated using Average Split which computes average accuracy from percentage split. The results reveal the fact that classification with optimally selected genes yields improved accuracy than classifying cancer subtypes with full gene expression data. It was identified that the optimal set of predictive genes will improve accuracy of classification while the reduction in gene number will aid towards reducing execution time. The Chi-Square test based serial and parallel classification approach reported improved accuracy compared with previously reported accuracy from AI Orange labs. Similarly Chi-Square test based serial and parallel classification yields improved accuracy compared with Rank Weight Feature Selection based classification for brain tumor, lung cancer with 3 class and childhood Leukemia with 4 classes. In future, this research work is directed towards expanding the application of the proposed parallel gene selection algorithm to Gastric tumor and Glioblastoma datasets. Moreover, detecting the underlying cause for reduced accuracy during parallel execution is also the focus of this research. The authors propose to further investigate the possibility of applying the proposed parallel algorithms to other large clinical datasets like Protein53 (P53) and determine whether parallelized classification accuracy can be enhanced sans sacrificing execution time.

Key Points

1. Microarray Gene Expression data is low sample and high dimensional data that necessitates pre-processing and feature (gene) reduction.
2. To scale huge Microarray Gene data, it is essential to parallelize data mining algorithms (Gene Selection and Classification) using Apache Spark framework.
3. Parallelized classification followed by parallelized gene selection improves accuracy and reduces execution time.
4. 10-fold Average Split accuracy has been used to evaluate the constructed model. Parallelized Chi-Square test based classification has been compared with previously reported accuracy from AI Orange labs Ljubljana.
5. Execution of Lung cancer dataset with 5 diagnostic classes failed with serial execution, whereas it is made feasible with parallelized algorithms on Spark Machine Learning Library (SparkMLlib).

References

1. Jacob, S.G., and R.G. Ramani. 2012. Data mining in clinical data sets: a review training. *International Journal of Applied Information Systems* 4 (6): 15–26.
2. Piatetsky-Shapiro, G., and P. Tamayo. 2003. Microarray data mining: Facing the challenges. *ACM SIGKDD Explorations Newsletter* 5 (2): 1–5.
3. Golub, T.R., D.K. Slonim, P. Tamayo, et al. 1999. Molecular classification of cancer: class discovery and class prediction by gene expression monitoring. Science 286 (5439): 531–537.
4. Liu, H., R.G. Sadygov, and J.R. Yates. 2004. A model for random sampling and estimation of relative protein abundance in shotgun proteomics. *Analytical Chemistry* 76 (14): 4193–4201.
5. Helleputte, T., and P. Dupont. 2009. Feature selection by transfer learning with linear regularized models. In *Joint European conference on machine learning and knowledge discovery in databases*, 533–547. Berlin Heidelberg: Springer.
6. Guyon, I., and A. Elisseeff. 2003. An introduction to variable and feature selection. *Journal of Machine Learning Research*: 1157–1182.
7. Guan, P., D. Huang, M. He, et al. 2009. Lung cancer gene expression database analysis incorporating prior knowledge with support vector machine-based classification method. *Journal of Experimental and Clinical Cancer Research* 28 (1): 1–7.
8. Rangarajan, L. 2010. Bi-level dimensionality reduction methods using feature selection and feature extraction. *International Journal of Computer Applications.* 4 (2): 33–38.
9. Gracia Jacob, S. 2015. Discovery of novel oncogenic patterns using hybrid feature selection and rule mining. Ph.D. thesis. Anna University.
10. Han, J., and Micheline, Kamber. 2006. *Data mining concepts and techniques*, 2nd ed. Elsevier.
11. Jirapech-Umpai, T., and S. Aitken. 2005. Feature selection and classification for microarray data analysis: Evolutionary methods for identifying predictive genes. *BMC Bioinformatics* 6 (1): 1–11.

12. Masih, S., and S. Tanwani. 2014. Data mining techniques in parallel and distributed environment-a comprehensive survey. *International Journal of Emerging Technology and Advanced Engineering* 4 (3): 453–461.
13. Pakize, S.R., and A. Gandomi. 2014. Comparative study of classification algorithms based on MapReduce model. *International Journal of Innovative Research in Advanced Engineering*: 2349–2163.
14. Parallel Programming Framework Apache Spark. http://spark.apache.org/. Accessed 9 Nov 2016.
15. Meng, X., J. Bradley, B. Yuvaz, et al. 2016. Mllib: Machine learning in apache spark. *Journal of Machine Learning Research*. 17 (34): 1–7.
16. Hall, M., E. Frank, G. Holmes, & I.H. Witten et al. 2009. The WEKA data mining software: an update. *ACM SIGKDD Explorations Newsletter* 11 (1): 10–18.
17. Parallel Programming Framework Spark. Machine Learning Library (SparkMLlib). http://spark.apache.org/docs/latest/mllib-guide.html. Accessed 6 Nov 2016.
18. Artificial Intelligence Orange Labs. Ljubljana. http://www.biolab.si/supp/bi-cancer/projections/. Accessed 31 Oct 2016.
19. Hall, M. 1999. Correlation-based feature selection for machine learning. Ph.D. thesis.
20. Kuncheva, L.I. 1992. Fuzzy rough sets: Application to feature selection. *Fuzzy Sets and Systems* 51 (2): 147–153.
21. Geng, X., T.Y. Liu, T. Qin et al. 2007. Feature selection for ranking. In *Proceedings of the 30th annual international ACM SIGIR conference on research and development in information retrieval*, 407–414.
22. Shannon, C.E. 2001. A mathematical theory of communication. *ACM SIGMOBILE Mobile Computing and Communications Review* 5 (1): 3–55.
23. Karegowda, A.G., A.S. Manjunath, and M.A. Jayaram. 2010. Comparative study of attribute selection using gain ratio and correlation based feature selection. *International Journal of Information Technology and Knowledge Management* 2 (2): 271–277.
24. Jiang, B.N., X.Q. Ding, L.T. Ma, et al. 2008. A hybrid feature selection algorithm: Combination of symmetrical uncertainty and genetic algorithms. In *The second international symposium on optimization and systems biology*, 152–157.
25. Forman, G. 2003. An extensive empirical study of feature selection metrics for text classification. *Journal of Machine Learning Research*: 1289–305.
26. Kira, K., and L.A. Rendell. 1992. A practical approach to feature selection. In *Proceedings of the ninth international workshop on Machine learning*, 249–256.
27. Alonso-González, C.J., Q.I. Moro-Sancho, A. Simon-Hurtado, et al. 2012. Microarray gene expression classification with few genes: Criteria to combine attribute selection and classification methods. *Expert Systems with Applications* 39 (8): 7270–7280.
28. Zhang, H., L. Li, C. Luo, et al. 2014. Informative gene selection and direct classification of tumor based on chi-square test of pairwise gene interactions. *BioMed Research International* 2014: 1–10.
29. Begum, S., D. Chakraborty, and R. Sarkar. 2015. Cancer classification from gene expression based microarray data using SVM ensemble. In *2015 International conference on condition assessment techniques in electrical systems (CATCON)*, 13–16. IEEE.
30. Jeyachidra, J., and M. Punithavalli. 2013. A comparative analysis of feature selection algorithms on classification of gene microarray dataset. In *Information communication and embedded systems (ICICES), IEEE 2013 international conference on 2013*, 1088–1093.
31. Weitschek, E., G. Fiscon, G. Felici, et al. 2015. Gela: A software tool for the analysis of gene expression data. In *2015 26th international workshop on database and expert systems applications (DEXA) IEEE*, 31–35.
32. Cabrera, J., A. Dionisio, G. Solano. 2015. Lung cancer classification tool using microarray data and support vector machines. In *Information, Intelligence, Systems and Applications (IISA), 2015 6th International Conference*. IEEE, 1–6.
33. Nguyen, C., Y. Wang, and H.N. Nguyen. 2013. Random forest classifier combined with feature selection for breast cancer diagnosis and prognostic. *Journal of Biomedical Science and Engineering*. 6 (5): 551–560.

34. Rajeswari, K., V. Vaithiyanathan, and S.V. Pede. 2013. Feature selection for classification in medical data mining. *International Journal of Emerging Trends and Technology in Computer Science (IJETTCS).* 2 (2): 492–497.
35. Lavanya, D., and K.U. Rani. 2012. Ensemble decision tree classifier for breast cancer data. *International Journal of Information Technology Convergence and Services.* 2 (1): 17–24.
36. Ben-Dor, A., L. Bruhn, N. Friedman, et al. 2000. Tissue classification with gene expression profiles. *Journal of Computational Biology* 7 (3–4): 559–583.
37. Hassanien, A.E. 2003. Classification and feature selection of breast cancer data based on decision tree algorithm. *Studies in Informatics and Control.* 12 (1): 33–40.
38. Kashyap, H., H.A. Ahmed, N. Hoque, et al. 2015. Big data analytics in bioinformatics: A machine learning perspective. arXiv preprint arXiv:1506.05101. 13 (9): 1–20.
39. Stokes, T.H., R.A. Moffitt, J.H. Phan, et al. 2007. chip artifact CORRECTion (caCORRECT): a bioinformatics system for quality assurance of genomics and proteomics array data. *Annals of Biomedical Engineering* 35 (6): 1068–1080.
40. Phan, J.H., A.N. Young, and M.D. Wang. 2013. omniBiomarker: a web-based application for knowledge-driven biomarker identification. *IEEE Transactions on Biomedical Engineering* 60 (12): 3364–3367.
41. Li. M., J. Tan, Y. Wang, et al. 2015. Sparkbench: A comprehensive benchmarking suite for in memory data analytic platform spark. In *Proceedings of the 12th ACM international conference on computing frontiers,* vol. 53, 1–8.
42. Koliopoulos, A.K., P. Yiapanis, F. Tekiner, et. al. A parallel distributed weka framework for big data mining using spark. In *2015 IEEE international congress on big data,* 9–16.
43. Shafer, J., R. Agrawal, and M. Mehta. 1996. SPRINT: A scalable parallel classifier for data mining. In *Proceeding of the 1996 international conference,* 544–555. Very Large Data Bases.
44. Chauhan, H., and A. Chauhan. 2013. Implementation of decision tree algorithm c4. *International Journal of Scientific and Research Publications* 3 (10): 1–3.
45. Wakayama, R., R. Murata, A. Kimura, et al. 2015. Distributed forests for MapReduce-based machine learning. In *Proceedings of the IAPR Asian conference on pattern recognition (ACPR),* 1–5.
46. Han, J., Y. Liu, and X. Sun. A scalable random forest algorithm based on MapReduce. In *Software engineering and service science (ICSESS), 2013 4th IEEE international conference on 2013,* 849–852.
47. Li, B., X. Chen, M. J. Li, et al. 2012. Scalable random forests for massive data. In *Pacific-Asia conference on knowledge discovery and data mining,* 135–146. Springer Berlin Heidelberg.
48. Hall, L.O., N. Chawla, and K.W. Bowyer. 1998. Combining decision trees learned in parallel. In *Working notes of the KDD-97 workshop on distributed data mining,* 10–15.
49. Amado, N., J. Gama, and F. Silva. 2004. Exploiting parallelism in decision tree induction. In *Proceedings from the ECML/PKDD workshop on parallel and distributed computing for machine learning,* 13–22.
50. Richards JW, Eads D, Bloom JS, Brink H, Starr D. WiseRFTM: A fast and scalable Random Forest. A WHITE PAPER from wise.io. 2013.
51. Islam, A.T., B.S. Jeong, A.G. Bari, et al. 2015. MapReduce based parallel gene selection method. *Applied Intelligence* 42 (2): 147–156.
52. Peralta, D., S. del Río, S. Ramírez-Gallego, et al. 2015. Evolutionary feature selection for big data classification: A mapreduce approach. *Mathematical Problems in Engineering* 2015: 1–11.
53. Wang, X., and O. Gotoh. 2010. A robust gene selection method for microarray-based cancer classification. *Cancer Informatics* 9: 15–30.
54. Wu, G., H. Li, X. Hu, et al. 2009. MReC4. 5: C4. 5 ensemble classification with MapReduce. In *2009 fourth ChinaGrid annual conference,* 249–255. IEEE.
55. Wu, Z., Y. Li, A. Plaza, et al. 2016. Parallel and distributed dimensionality reduction of hyperspectral data on cloud computing architectures. *IEEE Journal of Selected Topics in Applied Earth Observations and Remote Sensing* 9 (6): 2270–2278.

56. Ramani, R.G., and S.G. Jacob. 2013. Benchmarking classification models for cancer prediction from gene expression data: A novel approach and new findings. *Studies Informatics Control* 22 (2): 134–143.

57. Das, H., B. Naik, and H.S. Behera. 2018. Classification of diabetes mellitus disease (DMD): A data mining (DM) approach. In *Progress in computing, analytics and networking*, 539–549. Singapore: Springer.

58. Das, H., A.K. Jena, J. Nayak, B. Naik, and H.S. Behera. 2015. A novel PSO based back propagation learning-MLP (PSO-BP-MLP) for classification. In *Computational intelligence in data mining*, vol. 2, 461–471. New Delhi: Springer.

59. Sahoo, A.K., S. Mallik, C. Pradhan, B.S. Mishra, R.K. Barik, and H. Das. 2019. Intelligence-based health recommendation system using big data analytics. In *In big data analytics for intelligent healthcare management*, 227–246. Academic Press.

60. Dey, N., H. Das, B. Naik, & H.S. Behera (Eds.). 2019. *Big data analytics for intelligent healthcare management*. Academic Press.

An Evolutionary Algorithm Based Hybrid Parallel Framework for Asia Foreign Exchange Rate Prediction

Minakhi Rout and Koffi Mawuna Koudjonou

Abstract This paper is about an evolutionary algorithm-based hybrid parallel model to enhance the prediction of Asia foreign exchange rate. This hybrid parallel model is made up of a trained adaptive linear combiner as linear model and a functional link artificial neural network (FLANN) as a non-linear model in parallel. To set the parameters of the non-linear model, differential evolution (DE) learning algorithm has been employed whereas the linear model has already been trained using LMS algorithm. We have primarily focused on Asia foreign exchange rate prediction for which, we have considered six Forex data set to validate the model and the study reveals that it outperforms than other models.

Keywords Asia foreign exchange rate · LMS based linear model · Differential evolution (DE) · Functional link artificial neural network (FLANN)

1 Introduction

An exchange rate is the value of one currency expressed in terms of another currency. Whenever one country's currency rate goes through depreciation and appreciation condition, the exchange rate forecasting helps manage the future exchange rate risks. Exchange rate may be required by companies to hedge against potential losses, arranging short-term and long-term funds and for performing investments and to assess achieves of a foreign subsidiary. The quality of decisions, in such cases, depends on the accuracy of exchange rate projections. Nowadays, every company or organization which deals with foreign companies mostly requires exchange rate forecasting to predict future exchange rate values, to minimize the exchange risks and to maximize the returns of companies. The exchange rate forecasting is also necessary to evaluate the foreign denominate cash flows involved in international

M. Rout (✉) · K. M. Koudjonou
School of Computer Engineering, KIIT Deemed to be University, Bhubaneswar, Odisha, India
e-mail: minakhi.rout@gmail.com

K. M. Koudjonou
e-mail: koudjonoukoffi@gmail.com

© Springer Nature Switzerland AG 2020
M. Rout et al. (eds.), *Nature Inspired Computing for Data Science*,
Studies in Computational Intelligence 871,
https://doi.org/10.1007/978-3-030-33820-6_11

279

transactions. Hence, exchange rate prediction is very important to evaluate the benefits and risks attached to the international business environment. As a result, it is important that the limitations of the forecasts are understood so they can be used as effectively as possible. The objective of the study is to improve the exchange rate prediction accuracy with respect to the model used in [1] in the same domain, we have motivated to use differential evolution based learning algorithm to optimize the weights of nonlinear low complexity neural network model i.e. FLANN so that it can converge faster. From the related work section, we have observed that differential evolution optimization algorithm is widely used in the domain of financial forecasting and converge faster than other optimization techniques. Novelty of this work is to hybridize the Model with Differential evolution optimization and utilizing two other statistical features kurtosis and skewness. Extensive study has done with the six different Asia and Pacific forex dataset to validate the model.

The remaining parts of this paper are organized as follows, Sect. 2 gives the extensive overview of related works and motivation towards this research work. Section 3 describes the Differential Evolution, the evolutionary algorithm in details. The proposed hybrid model and its working principle presented in Sect. 4. Section 5 elaborates the details of data sets and the preparation of input feature sets followed by simulation results and discussion in the succeeding Sect. 6. Finally, the work has been concluded in Sect. 7.

2   Related Works

Foreign exchange rate forecasting is the one of the most vital challenge for many companies, especially for those in finance sector. For dealing with such challenge, many forecasting models are available. These models are often basically adaptive and are build using specific learning algorithms during training. In foreign exchange rate prediction, accurate prediction of exchange rate is the most important part for various financial institutes and currency exchange market because exchange rate data contains uncertain and nonlinear data series and this exchange rate prediction can have a very deep impact on a whole country's both economic and political condition. Thus, its accurate prediction is a very crucial task. ARMA (autoregressive moving average) combined with differential evolution (DE) is one of the models used for solving the complex forecasting of exchange rate predictions. It is a hybrid prediction model in which a dataset of previous time series values is used to train the model by updating its parameters using DE optimization algorithm [2]. To reduce the risk of investment in mutual funds, FLANN (Functional Link Artificial Neural Network) which is a single layer and single neuron architecture with nonlinear input is used as the adaptive model for NAV (net asset value prediction) [3]. This model is noteworthy for its fast convergence, less computational complexity and accurate results. Functional link artificial neural network is also very useful when it comes to classification task [4]. It is used to solve the tough task which consist of finding the optimal nonlinear boundary for classification problem in data mining by taking advantage of the nonlinear learning capabilities of artificial neural networks, an effective method to design many complex applications like functional approximation, nonlinear system identification and control, supervised classification and optimization. In [5], the behavior of Indian Foreign exchange markets using autoregressive integrated moving average (ARIMA), neural network and fuzzy models have been explained. For constructing an ARIMA model for exchange rate time series, the autoregressive, moving average and the integrated term successively represented by (p, q, d) are required parameters. To identify integrated auto regressive term and moving average term, autocorrelation (ACF) and partial auto correlation (PACF) are used.

Reference [6] discusses about the foreign exchange Random Walk Hypothesis (RWH). RWH states that market prices change with respect to random walk or the change in random way, which make their prediction a complicated task. Another work is related to neural evolutionary algorithm based on Cartesian genetic programming ANN (CGPANN) and Recurrent Cartesian genetic programming ANN (RCGPANN) model. The RCGPANN models work through feedback method in the network by feeding back one or more estimated outputs into the input of the system [7]. RMB exchange rate used back propagation (BP) neural network to convert the Chinese currency to another currency. In the work, the neural network model has two parts: signal propagation and error signal reverse dissemination. In first part, weighted input goes through input layer, then the output of input layer operates in hidden layer and passes through output layer and error signal reverse dissemination to obtain the error signal of each units of revised weight value. The process is called weight continual readjustment process [8]. Nowadays, the accurate prediction of stock price indices of interest is also challenging topic. To overcome it, [9] proposed a hybrid prediction model composed by Radial basis Function (RBF) neural network and non-dominated sorting multi-objective genetic algorithm-II (NSGA). Next, feed forward and back propagation neural network is used to solve exchange rate forecasting with minimum error where BP algorithm makes use of the distinct passes. In the first (forward) pass, the input vector to the first layer is propagated forward through the network, and the second (backward) pass attempts to correct the error in it [10]. Some traditional forecasting models are used for forecasting seasonal time series (i.e. fluctuation duration at least 2 years), but these are not convenient for all. To overcome of this issue, the following research paper gives the idea about how the FLANN model is used for solving problem of seasonal time series forecasting (TSF). To do so, FLANN model takes the row data, processes it and compares it with Random walk model and Feed forward neural network to get the accurate results [11]. The work in [12] is about the utilization of Differential Evolution (DE) algorithm for training the Feed Forward Flat Neural Network for classification of Parity-P problem. It uses Error Back propagation algorithm, evolutionary EA-NNT method and Levenberg Marquardt algorithm and compares the result obtained by Differential Evolutionary (DE) algorithm. The following is a differential evolution based functional link artificial neural Network (FLANN) model used to predict the Indian stock market indices. During the training, the output is taken form FLANN network and optimized with back propagation (BP) training algorithm first and then differential evolutionary (DE) algorithm to build two different models. Later, the results obtained by training the FLANN with least mean square algorithm are compared with those obtained by using differential evolution algorithm [13]. Time series forecasting (TFS) is an important tool in exchange rate forecasting, notably where traditional TSF process (i.e. statistical methods) cannot give accurate result. For instance, [14] describes how evolutionary artificial neural network (i.e. training using evolutionary algorithm) is used for TSF to get accurate forecasting. Two evolutionary algorithms: differential evolutionary (DE) algorithm and genetic algorithm are used for better result. The work done in [15] is about the methodology carried out by the automatic design of ANN to overcome the forecasting of the time series problem. Here two steps are used to solve the TSF problem. The first step explains about all settings of ANN for an efficient forecasting task using MPL (Multilayer Perceptron) and BP (Back Propagation) for learning. The second step explains about all parameters related to ANN such as input nodes, number of hidden layer, learning rate for BP along with how the overall architecture carry out search process performances by differential evolutionary (DE) algorithm and genetic algorithm. Guided by the wish of predicting a more accurate result, the authors of [16] combine different machine learning models in order to benefit from the advantage of each of them taken separately. The overall model consists of three different models. One of them is AMA (Adaptive Moving Average), a single neuron architecture with multiple input trained by least mean square algorithm to minimize the mean square error loss. Another model used is the AARMA (Adaptive Auto Regressive Moving Average), a combination of a forward pass (moving average) and backward pass (auto regressive) where the MA inputs are the samples features and the AR inputs are a fixed number of outputs got from the MA part. The MA and AR outputs are combined linearly to get the output of the overall AARMA model. Both the forward and backward neural network are trained using least mean square algorithm. The third model is an RBF (Radial Basis Function) neural network. All these models are trained individually and combined to form the ensemble model. The output of the three models are weighted and PSO algorithm is used to train the overall model. The results show better performance than the models taken apart. Similar work has been done in [17] which propose also an ensemble model but for exchange rate prediction. Using principal component analysis (PCA), this research sort out one of the most important question when it comes to choose the number of models to add together in the ensemble model, since we can't combine a randomly chosen number of individual models to have a better accuracy. Two models are used to reach the goal: a simple artificial neural network, a Generalized linear auto-regression (GLAR) and a combination of the previous two models.

3 Differential Evolution Algorithm

Differential evolution algorithm is a population based stochastic search introduced by Kennetn Price and Rainer Storn in 1995. It works like an efficient global optimizer in the continuous search domain. Differential evolution (DE) algorithms are developed to optimize real parameter valued function. It is mostly used in various fields like global numerical optimization, pattern recognition, image pixel clustering, text document clustering, large scale power dispatch problem and so on. Global optimization is necessary in fields such as engineering, statistics and finance. But many practical problems have objective functions that are non-differentiable, non-linear, noisy, flat, multi- dimensional, having many local minima or stochasticity. Such problems are difficult, if not impossible to solve analytically. Differential evolutionary can be used to find approximate solution to such problem.

For instance, the general problem formulation is $f : \sqsubseteq R^D \rightarrow R$ where the feasible region $X \neq 0$, the minimization problem is exposed as: $x^* \in X$ Such that $f(x^*) \leq f(x) \forall x \in X, f(x^*) \neq -\infty$.

The steps of differential evolution algorithm are the following.

3.1 Initialization

DE generates an initial population size of N and D-dimensional vector represented as $X_{i,G} = [x_{1,i,G}, x_{2,i,G} \ldots x_{D,i,G}]$, where $i = 1, 2, \ldots N$ is total population and G is generation number. Each parameter value is selected randomly and uniformly between upper and lower parameter bound x_j^L, x_j^U and represented as $x_j^L \leq x_{i,j,1} \leq x_j^U$.

The initial value of the jth parameter in case of jth individual at generation $G = 0$ is generated by

$$x_{i,0}^j = x_j^u + rand\,(0, 1) * (x_j^U - x_j^L) \tag{1}$$

where $rand\,(0, 1)$ represents uniformly distributed random number laying between 0 and 1, $j = 1, 2, \ldots D$.

3.2 Mutation

Foreach of N parameter vector undergoes through mutation operation and produce a mutant vector $V_{i,G}$ such as $V_{i,G} = [V_{i,G}^1 \ldots V_{i,G}^D]$. For each target parameter vector $x_{i,G}$ is computed as

$$v_{i,G+1} = x_{r1,G} + F(x_{r2,G} - x_{43,G}) \tag{2}$$

where F is mutation factor which value lies between 0 and 2 and $x_{r1,G}, x_{r2,G}, x_{r3,G}$ are randomly selected three vectors for target parameter $x_{i,G}$, such that the indices r_1, r_2, r_3 are distinct.

3.3 Crossover

In crossover operation, a trial vector $u_{i,G+1}$ is developed from element of target vector $x_{i,G}$ and the element of the mutant vector $v_{i,G+1}$.

$$u_{j,i,G+1} = \begin{cases} v_{j,i,G+1} & if \ rand_{j,i} \leq CR \ or \ j = Irand \\ x_{j,i,G+1} & if \ rand_{j,i} > CR \ and \ j = Irand \end{cases} \tag{3}$$

where $i = 1, 2 \ldots N$; $j = 1, 2, \ldots D$ and CR is crossover rate. $Irand$ is a random integer from $[1, 2 \ldots D]$. The crossover operator is between the jth parameter of $v_{i,G}$ and the corresponding element in the trial vector $u_{i,G}$ if $(rand_j, i \leq CR)$ or $(j = Irand)$. Otherwise, it is copied from the corresponding target vector $x_{i,G}$ and $Irand$ ensures that $v_{i,G+1} \neq x_{i,G}$.

3.4 Selection

In selection operation, the function value of all target vector $f(x_{i,G})$ is compared with function value of its trail vector $f(v_{i,G+1})$ and the one with the lowest function values is admitted to the next generation.

$$x_{i,G+1} = \begin{cases} u_{i,G+1} & if \ f(u_{i,G+1}) \leq f(x_{i,G}) \\ x_{i,G} & Otherwise \end{cases} \tag{4}$$

where $I = 1, 2 \ldots N$.

The above operations are repeated until a stopping criterion is reached.

4 Proposed Evolutionary Based Hybrid Forecasting Model

This model is a hybrid evolutionary based forecasting model designed for currency exchange rate forecasting. It consists of three stages. In the first stage, the extracted input training patterns are given to the simple adaptive linear combiner (ALC) to obtain the forecast price and the error, which is calculated based on available desired output. Then the parameters of the model are updated using least mean square learning algorithm. This process continues for each of the training pattern till the minimum squared error converged. Once the squared error converged then the weights of the linear combiner are frozen. At the second stage, the input training patterns are given to functional link artificial neural network (FLANN) in addition to the adaptive linear combiner in parallel to obtain the forecast value. At this stage, the parameters of the FLANN are getting updated whereas parameters of the linear combiner remain fixed.

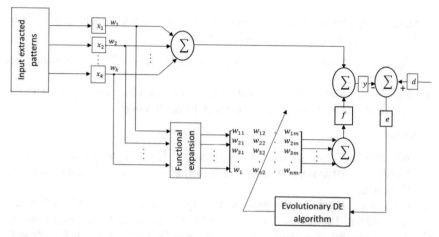

Fig. 1 Detailed block diagram of proposed evolutionary based hybrid forecasting model

The linear combiner can be considered as a guide who helps the FLANN to improve the performance. The third stage emerges from the second stage. To enhance the prediction of this parallel hybrid model, we have proposed an evolutionary based learning algorithm to set the parameters of the FLANN in second stage as depicted in Fig. 1.

The details about the working principle of the proposed model stepwise are explained as follow.

4.1 Phase I: Training of the ALC Model

i.   Initialize the weights of the of ALC randomly and set the learning parameter alpha between [0,1];
ii. Repeat steps iii to v to for each input training pattern;
iii. Obtained the output of the ALC model;
iv. calculate the error;
v.   update the weight using LMS learning algorithm;
vi. Repeat the above process till the squared error reached at a minimum level;
vii. freeze the weights of the ALC model for use in Stage II.

Phase II

i. Apply the input training patterns to the trained ALC model using the frozen weights obtained from Stage I to get its output;
ii. Using trigonometric expansion function, functionally expand the input elements to training patterns non-linearly as follows;

Input elements :$\{x_1, x_2,...x_n\}$

Functionally expanded training pattern for the above input elements:$\{x_1, sin\Pi x_1, cos\Pi x_1,, x_2, sin\Pi x_2, cos\Pi x_2, x_3, sin\Pi x_3, cos\Pi x_3\}$

iii. Apply the functionally expanded input training patterns to FLANN model to get its output;
iv. Finally, the output of the parallel model is obtained by summing up both the output of linear combiner and non-linear FLANN model.

Phase III

i.   Initialize randomly parameters (weights and bias) of the non-linear model FLANN as initial population (target vector) of Differential Evolution learning algorithm;
ii.   Each of the input training patterns are applied to the model with each of the target vector. The error will be obtained to serve as fitness of each target vector;
iii. The parameters of the model (target vector of DE) will be updated as the procedure of DE algorithm;
iv.   The procedure will be repeated till the lease mean square error achieved.

The procedure of the proposed model has been detailed diagrammatically in Fig. 2.

5 Forex Data Set and Preparation of Feature Sets

In this paper, we have considered the Asia foreign currency exchange data set for the evaluation and analysis. The Forex data sets are collected from the website http://www.forecasts.org. These data sets are time series data set as the currency price is recorded on first day of each month. The historical data set needs to be processed in order to extract the essential features, which are used later as inputs to the model. In our study we have considered 6 Asia Forex datasets for extensive analysis. The details regarding the data sets are given in Table 1. For the preparation of feature set out of the raw historical data are discussed below and trends of these dataset are plotted in Fig. 3.

Steps for preparation of feature sets:

i.   Collection of historical Forex data
ii.   Normalize the data set by min-max normalization
iii. Divide the data set into varying size of sliding window size. In our study, we have used window size of 12. From each group of sliding windows, extract the input features. In this study, we have extracted the following features: (12th price, mean, standard deviation, kurtosis and skewness) and these extracted features from each group of the sliding window is used as inputs for the forecasting models.

Kurtosis is a measure of whether data are lightly-tailed or heavily tailed with respect to normal distribution.

Skewness is a measure of symmetry or lack of symmetry.

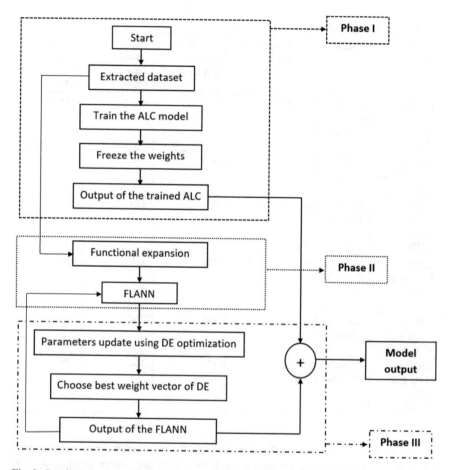

Fig. 2 Detailed diagram of the working procedure of the proposed model

6 Experimental Study and Result Analysis

The simulation is carried out using 6 different Asia currency exchange data sets. The evolutionary based proposed hybrid parallel model is implemented in MATLAB environment. In FLANN we have considered trigonometric expansion and we have expanded the input patterns with 3 expansions. The prediction is made using the proposed model and also the adaptive linear combiner (ALC), the functional link artificial neural network (FLANN) and the parallel combined (ALC-FLANN) model. The performance of the proposed model is obtained and compared with other three models in terms of mean absolute percentage error (MAPE) and root mean square error value (RMSE) and the results are listed in Table 2. From the Table 2, it is clearly visible that the DE based hybrid parallel model is performing better in all aspects that the other three models. The comparison of actual and forecast exchange

Table 1 Asia foreign exchange data set

Asia Forex data set	Date range	Total no. of data	Total input patterns extracted	No. of training sample (80%)	No. of testing samples (20%)
1US$ to Indian Rupees (INR)	01/01/1973 to 01/10/2017	538	526	421	105
1US$ to Japanese Yen (JPY)	01/01/1971to 13/10/2017	562	550	440	110
1US$ to Malaysia Ringgit (MYR)	01/01/1971to 13/10/2017	562	550	440	110
1US$ to Singapore Dollar (SGD)	01/01/1981to 01/10/2017	442	430	344	86
1US$ to China Yuan (CNY)	01/01/1981to 01/10/2017	442	430	344	86
1US$ to Hong kong Dollar (HKD)	01/01/1981to 01/10/2017	442	430	344	86

rate value of all the six data sets are carried out for one-month time horizon using the proposed model during training as well as testing and shown in Figs. 4, 5, 6, 7, 8 and 9. The figures depict that the forecast value matches the trend of original exchange rate value appropriately. Figures 10, 11 and 12 show the mean square error obtained during training using parallel (ALC-FLANN) model and proposed (ALC-FLANN-DE) model for IRN, SGD, and MRN data sets respectively.

7 Conclusion

Our work focused on an Evolutionary Algorithm based Hybrid Parallel Framework for exchange rate prediction. The study is carried out on six different Asia foreign exchange dataset for training the models and making predictions. The proposed model is constituted of an adaptive linear combiner (ALC) and a Functional Link Artificial Neural Network. The ALC is trained first and its weights are frozen. This trained model is then combined in parallel with a FLANN to form a hybrid model. The ALC guides during the training of the hybrid model which consist of updating

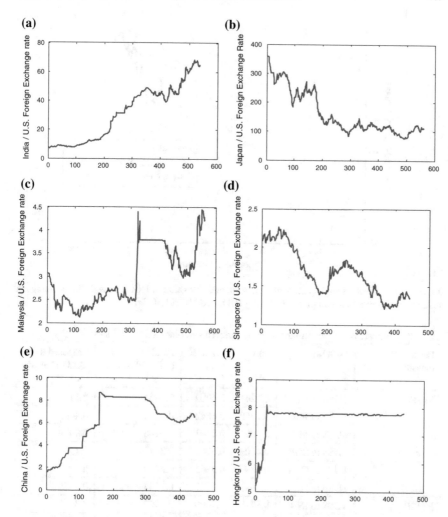

Fig. 3 Historical data and trends of foreign exchange data set as listed in Table 1

the parameters of the FLANN. Whereas the ALC parameters stay frozen over the training of the hybrid model, the parameters of the FLANN are updated using an evolutionary based learning algorithm. Afterward, using our ALC-FLANN-DE model, the prediction is made on the testing set of each dataset and based on the MAPE and RMSE, the results are compared with those of other models such as ALC, FLANN and ALC-FLANN. The simulation results reveal that the proposed model is performing better than the other three models. This is a step forward in the use of evolutionary algorithm combined with hybrid parallel models especially for time series data such as foreign exchange rate forecasting. If the ALC is not pre-trained properly then the model performance may be worse, is the limitation of the proposed model. This proposed model can also be used in the forecasting of other financial datasets. The

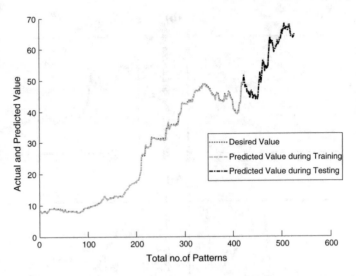

Fig. 4 Comparison of actual and predicted exchange rate value of INR dataset during training and testing for one month ahead prediction using DE based Hybrid Model

Table 2 Comparison of performance measure in terms of obtained MAPE and RMSE

Performance measure	Data set	ALC	FLANN	Parallel model (ALC-FLANN)	Proposed model (ALC-FLANN-DE)
MAPE	1US$ to INR	0.7241	0.6002	0.5864	0.5108
	1US$ to JPY	0.7569	0.6842	0.5268	0.4009
	1US$ to MYR	0.3891	0.3441	0.3246	0.2904
	1US$ to SGD	0.4003	0.3762	0.2472	0.1908
	1US$ to CNY	0.3892	0.3341	0.2109	0.1073
	1US$ to HKD	0.2643	0.1832	0.1045	0.0456
RMSE	1US$ to INR	0.5329	0.4090	0.3952	0.3196
	1US$ to JPY	0.8895	0.8168	0.6594	0.5335
	1US$ to MYR	0.1138	0.0688	0.0493	0.0151
	1US$ to SGD	0.2124	0.1883	0.0593	0.0029
	1US$ to CNY	0.2902	0.2351	0.1119	0.0083
	1US$ to HKD	0.2229	0.1418	0.0631	0.0042

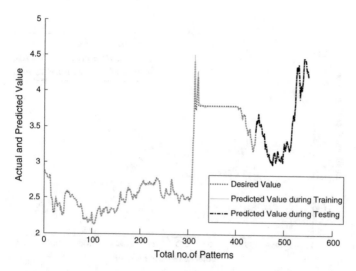

Fig. 5 Comparison of actual and predicted exchange rate value of JPY data set during training and testing for one month ahead prediction using DE based Hybrid Model

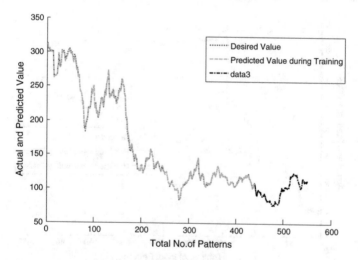

Fig. 6 Comparison of actual and predicted exchange rate value of MYR data set during training and testing for one month ahead prediction using DE based Hybrid Model

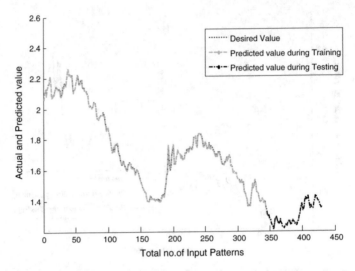

Fig. 7 Comparison of actual and predicted exchange rate value of SGD data set during training and testing for one month ahead prediction using DE based Hybrid Model

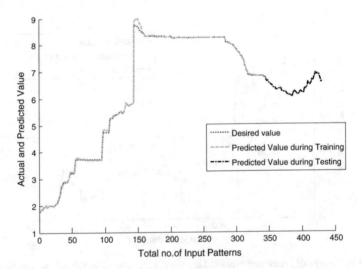

Fig. 8 Comparison of actual and predicted exchange rate value of CNY data set during training and testing for one month ahead prediction using DE based Hybrid Model

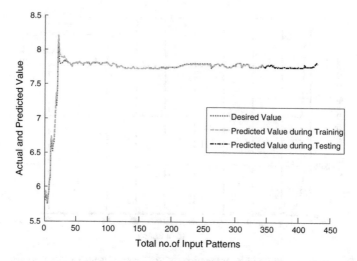

Fig. 9 Comparison of actual and predicted exchange rate value of HKD data set during training and testing for one month ahead prediction using DE based Hybrid Model

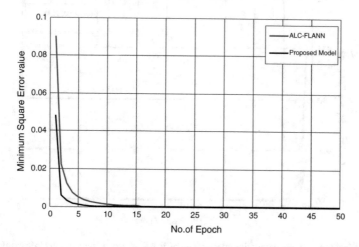

Fig. 10 Convergence characteristics comparison of INR dataset using various forecasting models

prediction accuracy can also be enhanced by incorporating some new and advanced optimization techniques such as JAYA, Social Group Optimization.

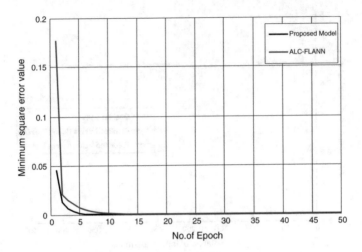

Fig. 11 Convergence characteristics comparison of SGD dataset using various forecasting models

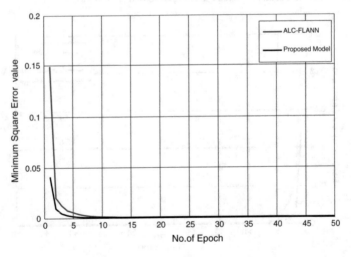

Fig. 12 Convergence characteristics comparison of MRN dataset using various forecasting models

References

1. Jena, P.R., R. Majhi, and B. Majhi. 2015. Development and performance evaluation of a novel knowledge guided artificial neural network (KGANN) model for exchange rate prediction. *Journal of King Saud University-Computer and Information Sciences* 27 (4): 450–457.
2. Rout, Minakhi, et al. 2014. Forecasting of currency exchange rates using an adaptive ARMA model with differential evolution based training. *Journal of King Saud University-Computer and Information Sciences* 26 (1): 7–18.

3. Anish, C. M., and Babita Majhi. 2016. Prediction of mutual fund net asset value using low complexity feedback neural network. In *Current Trends in Advanced Computing (ICCTAC), IEEE International Conference on.* IEEE.
4. Misra, B. B., and S. Dehuri. 2007. Functional link artificial neural network for classification task in data mining.
5. Babu, A., and S. Reddy. 2015. Exchange rate forecasting using ARIMA. *Neural Network and Fuzzy Neuron, Journal of Stock & Forex Trading.*
6. Bellgard, Christopher, and Peter Goldschmidt. 1999. Forecasting foreign exchange rates: Random Walk Hypothesis, linearity and data frequency. In *12th Annual Australasian Finance & Banking Conference.*
7. Rehman, Mehreen, Gul Muhammad Khan, and Sahibzada Ali Mahmud. 2014. Foreign currency exchange rates prediction using cgp and recurrent neural network. *IERI Procedia* 10: 239–244.
8. Majhi, Babita, Minakhi Rout, and Vikas Baghel. 2014. On the development and performance evaluation of a multiobjective GA-based RBF adaptive model for the prediction of stock indices. *Journal of King Saud University-Computer and Information Sciences* 26 (3): 319–331.
9. Ye, Sun. 2012. RMB exchange rate forecast approach based on BP neural network. *Physics Procedia* 33: 287–293.
10. Chen, Joseph C., and Naga Hrushikesh R. Narala. 2017. Forecasting Currency Exchange Rates via Feedforward Backpropagation Neural Network.
11. Slowik, Adam, and Michal Bialko. 2008. Training of artificial neural networks using differential evolution algorithm. *Human System Interactions, 2008 Conference on.* IEEE.
12. Mohapatra, Puspanjali, Alok Raj, and Tapas Kumar Patra. 2012. Indian stock market prediction using Differential Evolutionary Neural Network model. *International Journal of Electronics Communication and Computer Technology (IJECCT)* 2.
13. Panigrahi, Sibarama, YasobantaKarali, and H. S. Behera. 2013. Time series forecasting using evolutionary neural network. *International Journal of Computer Applications* 75 (10).
14. Donate, Juan Peralta, et al. 2013. Time series forecasting by evolving artificial neural networks with genetic algorithms, differential evolution and estimation of distribution algorithm. *Neural Computing and Applications* 22 (1): 11–20.
15. Divyapriya R., and R. ManickaChezhian. 2013. Accurate forecasting prediction of foreign exchange rate using neural network algorithms: A STUDY. *IJCSMC* 2 (7): 344–349.
16. Anish, C. M., Babita Majhi, and Ritanjali Majhi. 2016. Development and evaluation of novel forecasting adaptive ensemble model. *The Journal of Finance and Data Science* 2 (3): 188–201.
17. Yu, Lean, Shouyang Wang, and Kin Keung Lai. A novel nonlinear ensemble forecasting model incorporating GLAR and ANN for foreign exchange rates. *Computers & Operations Research* 32 (10): 2523–2541.

Correction to: Nature Inspired Computing for Data Science

Minakhi Rout, Jitendra Kumar Rout and Himansu Das

Correction to:
M. Rout et al. (eds.), *Nature Inspired Computing*
for Data Science, **Studies in Computational**
Intelligence 871,
https://doi.org/10.1007/978-3-030-33820-6

The original version of this book was published with an incorrect volume number, which has now been changed from "SCI 871" to "871" The correction book has been updated with the change.

The updated version of the book can be found at
https://doi.org/10.1007/978-3-030-33820-6

© Springer Nature Switzerland AG 2020
M. Rout et al. (eds.), *Nature Inspired Computing for Data Science*,
Studies in Computational Intelligence 871,
https://doi.org/10.1007/978-3-030-33820-6_12

Printed in the United States
By Bookmasters